CAMBRIDGE LIBRARY COLLECTION

Books of enduring scholarly value

Mathematical Sciences

From its pre-historic roots in simple counting to the algorithms powering modern desktop computers, from the genius of Archimedes to the genius of Einstein, advances in mathematical understanding and numerical techniques have been directly responsible for creating the modern world as we know it. This series will provide a library of the most influential publications and writers on mathematics in its broadest sense. As such, it will show not only the deep roots from which modern science and technology have grown, but also the astonishing breadth of application of mathematical techniques in the humanities and social sciences, and in everyday life.

Statics

A.S. Ramsey (1867-1954) was a distinguished Cambridge mathematician and President of Magdalene College. He wrote several textbooks 'for the use of higher divisions in schools and for first-year students at university'. This book on statics, published in 1934, was intended as a companion volume to his Dynamics of 1929 and like the latter was based upon his lectures to students of the mathematical tripos, but it assumes no prior knowledge of the subject, provides an introduction and offers more that 100 example problems with their solutions. Topics include vectors, forces acting at a point, moments, friction, centres of gravity, work and energy, and elasticity.

Cambridge University Press has long been a pioneer in the reissuing of out-of-print titles from its own backlist, producing digital reprints of books that are still sought after by scholars and students but could not be reprinted economically using traditional technology. The Cambridge Library Collection extends this activity to a wider range of books which are still of importance to researchers and professionals, either for the source material they contain, or as landmarks in the history of their academic discipline.

Drawing from the world-renowned collections in the Cambridge University Library, and guided by the advice of experts in each subject area, Cambridge University Press is using state-of-the-art scanning machines in its own Printing House to capture the content of each book selected for inclusion. The files are processed to give a consistently clear, crisp image, and the books finished to the high quality standard for which the Press is recognised around the world. The latest print-on-demand technology ensures that the books will remain available indefinitely, and that orders for single or multiple copies can quickly be supplied.

The Cambridge Library Collection will bring back to life books of enduring scholarly value across a wide range of disciplines in the humanities and social sciences and in science and technology.

Statics

A Text-Book for the use of the Higher Divisions in Schools and for First Year Students at the Universities

ARTHUR STANLEY RAMSEY

CAMBRIDGE
UNIVERSITY PRESS

CAMBRIDGE UNIVERSITY PRESS

Cambridge New York Melbourne Madrid Cape Town Singapore São Paolo Delhi

Published in the United States of America by Cambridge University Press, New York

www.cambridge.org
Information on this title: www.cambridge.org/9781108003155

This edition first published 1934
This digitally printed version 2009

ISBN 978-1-108-00315-5

STATICS

LONDON
Cambridge University Press
FETTER LANE

NEW YORK · TORONTO
BOMBAY · CALCUTTA · MADRAS
Macmillan

TOKYO
Maruzen Company Ltd

STATICS

A Text-Book for the use of the
Higher Divisions in Schools
and for
First Year Students at the Universities

by

A. S. RAMSEY, M.A.

President of Magdalene College
Cambridge

CAMBRIDGE
AT THE UNIVERSITY PRESS
1934

PRINTED IN GREAT BRITAIN

PREFACE

THIS book has been written as a companion volume to my book on *Dynamics* published a few years ago and is intended mainly for the same class of students, namely, for mathematical specialists in the higher divisions of schools and for students preparing for a degree in mathematics in the Universities. It is based in part upon courses of lectures given during many years to first-year students preparing for the Mathematical Tripos; and though many readers will already possess some knowledge of the subject, no such knowledge is assumed and an attempt has been made in the early chapters to present the subject in as simple a way as possible and with very detailed explanations.

The book deals with all those parts of the subject which are usually covered by the term Elementary Statics, with special attention to Graphical Statics, Friction and Virtual Work. For the use of more advanced students there are also chapters on the statics of flexible strings and the bending of rods, and the book concludes with a brief account of force systems in three dimensions.

There are nearly five hundred examples for solution taken mainly from papers set in either Scholarship, College, Intercollegiate or Tripos Examinations, and their sources are indicated by the letters S, C, I and T. More than a hundred examples are solved in the text, sometimes by alternative methods.

It has become the fashion of late to express mechanical relations in the symbolism of vector algebra and to use the methods of vector algebra in proving mechanical theorems. The method has its advantages; but I have not adopted it in this book, because to most readers it would represent a new technique. The effort necessary to acquire this technique would in many cases be a hindrance rather than a help to the grasping of the mechanical principles which the book is intended to teach, and should, I think, be made at a later stage. I have therefore made no use of vector analysis, but the way in which forces and couples obey the vector law of addition is fully explained, and the chapter on Vectors from the book on *Dynamics* appears here again in amplified form.

Readers who are familiar with the books of the late Dr Routh and of Sir Horace Lamb will realize something of my indebtedness to both these authors, but I am conscious of a greater debt than is apparent and I should like to take this opportunity of expressing my gratitude.

In conclusion I desire to thank the printers and readers of the University Press for their excellent work in setting up the book and eliminating mistakes, and in so far as the book still contains errors I shall be grateful to anyone who will point them out.

A. S. R.

November 1933
CAMBRIDGE

CONTENTS

Chapter I: INTRODUCTION

Chapter II: VECTORS

Chapter III: FORCES ACTING AT A POINT

Chapter IV: MOMENTS. PARALLEL FORCES. COUPLES

ART.		PAGE
4·5.	Centre of Parallel Forces	36
4·51.	Centre of Gravity	37
4·53.	Analytical Formulae for Centre of Parallel Forces .	38
4·6.	Couples	40
4·61–3.	Equivalence of Couples	41
4·64.	Specification of a Couple	44
4·65.	Composition of Couples	45
	Examples	50

Chapter V: COPLANAR FORCES

5·1.	Reduction to a Force at any Point and a Couple. .	56
5·2.	Conditions of Equilibrium	57
5·3.	Analytical Method	58
5·32.	Worked Examples	60
	Examples	63

Chapter VI: THE SOLUTION OF PROBLEMS

6·1.	Equations of Equilibrium	67
6·2.	Constraints and Degrees of Freedom	67
6·21.	Three Forces...Coplanar and Concurrent or Parallel .	68
6·3.	Problems of two or more Bodies	72
6·4.	Reactions at Joints	75
6·42.	Working Rules	77
6·5.	Chain of Heavy Particles	81
6·52.	Chain of Heavy Rods	83
	Examples	85

Chapter VII: BENDING MOMENTS

7·1, 7·2.	Stresses in a Beam	95
7·3.	Relations between Bending Moment and Shearing Force	98
7·4.	Worked Examples	100
	Examples	104

Chapter VIII: GRAPHICAL STATICS

8·1.	Graphical Determination of Resultant	107
8·2.	Pole of Force Diagram	109
8·3.	Parallel Forces	111
8·4.	Graphical Representation of Bending Moment . .	113
8·5.	Reciprocal Figures	116
8·6.	Frameworks	117

STATICS

Chapter I

INTRODUCTION

1·1. Statics is that branch of Mechanics which is concerned with the conditions under which bodies remain at rest relative to their surroundings. In such circumstances bodies are said to be in a state of equilibrium. It is assumed that they are acted upon by 'forces' which balance one another. The primary conceptions of Statics are forces and the bodies upon which they act. In Dynamics, force is defined as that which changes or tends to change the state of motion of a body, but in Statics we are not concerned with motion save in so far as a force would cause motion if unbalanced by another force. We get our ideas of force from the ability which we ourselves possess to cause, or to resist, the motion of our own bodies or of other bodies. We are conscious of measurable efforts required to lift bodies or to support bodies which would otherwise fall to the ground; we speak of these measurable efforts as forces which we exert, and we compare them with the 'weights' that we associate with bodies, by which term we mean 'the forces with which the Earth attracts them'. Thus the phrase 'a force of x pounds weight' means a force that would support a body weighing x pounds.

When holding the string of a kite flying in a gusty wind we are conscious of a pull or 'tension' on the string at the point where it leaves the hand that holds it, and we realize that this 'force' is something measurable and of varying measure, and that it acts now in this direction and now in that as the string moves hither and thither with the kite. This serves to illustrate the fact that a force possesses magnitude and acts in a definite line (that of the string) and may be regarded as acting at a definite point of that line (the point where the string leaves the hand). Thus in so far as a force possesses magnitude and direction it is a **vector** as defined in the next chapter, but whether forces obey the vector law of addition must be discussed later.

1·2. It is necessary to specify at the outset what it is upon which forces act. They cannot act upon nothing. Forces act upon material bodies, and unless the contrary is stated we shall assume that in every case the bodies are *rigid*, i.e. that the distance between each pair of particles that compose a body remains invariable. Actual bodies are all more or less elastic and capable of compression, extension or distortion under the action of forces, and the assumption of perfect rigidity is necessary in order to simplify the building up of the elementary theory of the subject.

We regard solid bodies as agglomerations of particles held together by forces of cohesion. We do not put any limit to the number, large or small, of particles that go to form a body, and there is no difficulty in the conception of a force acting upon a minute body or 'a single particle'.

1·3. The forces with which we shall be concerned are of three types: (*a*) a *push* or *thrust*; (*b*) a *pull* or *tension*; (*c*) an *attraction* such as the 'weight' of a body or 'the force with which the Earth attracts it'.

1·4. We shall assume (i) that if two equal and opposite forces act upon a body in the same straight line they have no effect upon the body's state of rest or motion, i.e. they balance one another. This is a statement the truth of which can easily be tested by experiment.

We shall also assume (ii) that the forces mutually exerted between two bodies are always equal and opposite. This is the *law of reaction* enunciated by Newton in the words 'Action and Reaction are equal and opposite'. It means that if a body A exerts a force F upon a body B, then B exerts an equal force F upon A but in the opposite direction.

1·5. A consequence of assumption (i) of **1·4** is what is sometimes called the *Principle of Transmissibility of Force*, viz. that a force may be supposed to act at *any* point in its line of action provided that the point is rigidly connected with the body on which the force acts.

For if P and Q are equal and opposite forces acting upon a body at the points A and B, then we may say that Q balances P no matter at what point B in the line AB the force Q acts; i.e. a force Q acting at A produces the 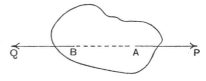 same effect as a force Q acting at B in the same straight line.

1·6. In the following chapter we shall discuss some of the common properties of a class of physical magnitudes called *vectors* and in Chapter III we shall give further reasons for including *force* in this class and then develop the consequences.

Chapter II

VECTORS

2·1. The physical quantities or measurable objects of reasoning in Applied Mathematics are of two classes. The one class, called **Vectors**, consists of all measurable objects of reasoning which possess directional properties, such as *displacement, velocity, acceleration, momentum, force,* etc. The other class, called Scalars, comprises measurable objects of reasoning which possess no directional properties, such as *mass, work, energy, temperature,* etc.

The simplest conception of a vector is associated with the displacement of a point. Thus the displacement of a point from A to B may be represented by the line AB, where the length, direction and sense (AB not BA) are all taken into account. Such a displacement is called a *vector* (Latin *veho*, I carry). A vector may be denoted by a single letter, e.g. as when we speak of 'the force P', or 'the acceleration f', or by naming the line, such as AB, which represents the vector. When it is desired to indicate that symbols denote vectors it is usual to *print* them in Clarendon type, e.g. **P**, and to *write* them with a bar above the symbol, e.g. $\overline{P}, \overline{AB}$.

Since the displacement from B to A is the opposite of a displacement from A to B, we write

$$\overrightarrow{BA} = -\overrightarrow{AB},$$

and take vectors in opposite senses to have opposite signs. Since two successive displacements of a point from A to B and from B to C produce the same result as a single displacement from A to C, we say that the vector AC is equal to the sum of the vectors AB, BC and write

$$\overline{AC} = \overline{AB} + \overline{BC} \ \dots\dots(1),$$

and further, if A, B, C, ... K, L are any set of points,

$$\overline{AL} = \overline{AB} + \overline{BC} + \dots + \overline{KL} \ \dots\dots\dots\dots(2).$$

Vectors in general are not localized; thus we may have a displacement of an assigned length in an assigned direction and sense but its locality not specified. In such a case all equal and parallel lines in the same sense will represent the same vector. On the other hand, vectors may be localized, either at a point, e.g. the *velocity* of a particle; or in a line, as for example a *force* whose line of action (but not point of application) is specified.

2·2. Composition of Vectors. A single vector which is equivalent to two or more vectors is called their **resultant**, and they are called the **components** of the resultant. Vectors are compounded by geometrical addition as indicated in (1) and (2) of the preceding Article.

A vector can be resolved into two components in assigned directions in any plane which contains the vector; for if AC be the vector, and through A, C two lines are drawn in the assigned directions meeting in B, then AB, BC are the components required.

When a vector is resolved into two components in directions at right angles to one another, each component is called the **resolved part of the vector** in the direction specified. Thus if a vector **P** makes an angle α with a given direction Ox, the resolved parts of **P** in the direction Ox and in the perpendicular direction Oy are

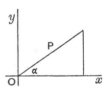

$$\mathbf{P}\cos\alpha \quad \text{and} \quad \mathbf{P}\sin\alpha.$$

Further, if Ox, Oy, Oz are three lines mutually at right angles, and the line OF represents a vector **P**, if we construct a rectangular parallelepiped with OF as diagonal and edges OA, OB, OC along Ox, Oy, Oz, as in the figure, then

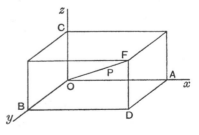

$$\overline{OF} = \overline{OA} + \overline{AD} + \overline{DF}$$

or $\quad \mathbf{P} = \overline{OA} + \overline{OB} + \overline{OC}.$

And, if **P** makes angles α, β, γ with Ox, Oy, Oz, we have, since OAF is a right angle, $\overline{OA} = \mathbf{P}\cos\alpha$, and similarly $\overline{OB} = \mathbf{P}\cos\beta$

and $\overline{OC} = \mathbf{P}\cos\gamma$, so that in this three-dimensional resolution of a vector its resolved parts in the three mutually perpendicular directions are

$$\mathbf{P}\cos\alpha, \quad \mathbf{P}\cos\beta, \quad \mathbf{P}\cos\gamma.$$

It is clear therefore that the resolved part of a vector along a given line is the orthogonal projection of the vector upon that line.

2·3. Let $\overline{AB}$, $\overline{BC}$, $\overline{CD}$, ... $\overline{KL}$ be a set of vectors forming sides of a polygon. Their resultant is the vector $\overline{AL}$ which completes the polygon. Let $a, b, c, d, \ldots k, l$ be the orthogonal projections of the points $A, B, C, D, \ldots K, L$ on any straight line Ox. Then, with due regard to signs,

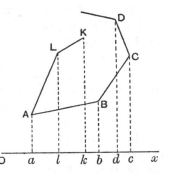

$$ab + bc + cd + \ldots + kl = al.$$

But these projections ab, bc, ... are the resolved parts of the vectors $\overline{AB}$, $\overline{BC}$, ... in the direction Ox, therefore the algebraical sum of the resolved parts of a set of vectors in an assigned direction is equal to the resolved part of their resultant in the same direction.

2·4. Analytical Method. To compound n vectors

$$\mathbf{P}_1, \mathbf{P}_2, \ldots \mathbf{P}_n.$$

(i) When the vectors are in the same plane.

Let the vectors make angles α_1, α_2, ... α_n with an axis Ox. Each vector may be resolved into two components, one in the direction Ox and the other in the perpendicular direction Oy. The components in direction Ox are equivalent to a single vector

$$\mathbf{X} = \mathbf{P}_1\cos\alpha_1 + \mathbf{P}_2\cos\alpha_2 + \ldots + \mathbf{P}_n\cos\alpha_n = \Sigma\,(\mathbf{P}\cos\alpha),$$

and the components in direction Oy are equivalent to a single vector

$$\mathbf{Y} = \mathbf{P}_1\sin\alpha_1 + \mathbf{P}_2\sin\alpha_2 + \ldots + \mathbf{P}_n\sin\alpha_n = \Sigma\,(\mathbf{P}\sin\alpha).$$

The two vectors $\mathbf{X}$, $\mathbf{Y}$ can now be compounded into a single vector $\mathbf{R}$ making an angle θ with Ox, such that

$$\mathbf{R}\cos\theta = \mathbf{X} \quad \text{and} \quad \mathbf{R}\sin\theta = \mathbf{Y},$$

and therefore

$$R^2 = X^2 + Y^2 \quad \text{and} \quad \tan\theta = Y/X \quad \text{.........(1)}.$$

(ii) When the vectors are not all in the same plane.

As in 2·2 take three axes Ox, Oy, Oz mutually at right angles and let the vectors make angles $\alpha_1, \alpha_2, \ldots \alpha_n$ with Ox, $\beta_1, \beta_2, \ldots \beta_n$ with Oy and $\gamma_1, \gamma_2, \ldots \gamma_n$ with Oz. Each vector may then be resolved into components of the types

$$\mathbf{P}\cos\alpha, \quad \mathbf{P}\cos\beta, \quad \mathbf{P}\cos\gamma$$

in the directions Ox, Oy, Oz. The components in direction Ox are equivalent to a single vector

$$\mathbf{X} = \mathbf{P}_1\cos\alpha_1 + \mathbf{P}_2\cos\alpha_2 + \ldots + \mathbf{P}_n\cos\alpha_n = \Sigma\,(\mathbf{P}\cos\alpha),$$

similarly the components in directions Oy and Oz are equivalent to single vectors

$$\mathbf{Y} = \mathbf{P}_1\cos\beta_1 + \mathbf{P}_2\cos\beta_2 + \ldots + \mathbf{P}_n\cos\beta_n = \Sigma\,(\mathbf{P}\cos\beta)$$

and

$$\mathbf{Z} = \mathbf{P}_1\cos\gamma_1 + \mathbf{P}_2\cos\gamma_2 + \ldots + \mathbf{P}_n\cos\gamma_n = \Sigma\,(\mathbf{P}\cos\gamma).$$

The three vectors $\mathbf{X}$, $\mathbf{Y}$, $\mathbf{Z}$ can now be compounded into a single vector $\mathbf{R}$ making angles θ, ϕ, ψ with Ox, Oy, Oz, such that

$$\mathbf{R}\cos\theta = \mathbf{X}, \quad \mathbf{R}\cos\phi = \mathbf{Y} \quad \text{and} \quad \mathbf{R}\cos\psi = \mathbf{Z} \quad \text{...(2)},$$

and by squaring and adding

$$R^2 = X^2 + Y^2 + Z^2 \quad \text{.................(3)}.$$

When the magnitude of R has been found from (3) its direction is determined by (2).

In obtaining (3) we have assumed that $\cos^2\theta + \cos^2\phi + \cos^2\psi = 1$; that this is true is seen from the figure of 2·2, where θ, ϕ, ψ may denote the inclinations of OF to the axes, then

$$\cos^2\theta + \cos^2\phi + \cos^2\psi = \frac{OA^2}{OF^2} + \frac{OB^2}{OF^2} + \frac{OC^2}{OF^2} = 1.$$

2·41. The method of obtaining the resultant in **2·4** is based on the fact that if the vectors are all resolved in any assigned direction Ox, then the resolved part of the resultant in that direction is equal to the algebraical sum of the resolved parts of the given vectors.

When the vectors are not all in the same plane each vector is resolved into three components in the directions of three rectangular axes

Ox, Oy, Oz chosen arbitrarily, so that for any direction Ox in space the resolved part of the resultant is equal to the algebraical sum of the resolved parts of the given vectors. Also, if each vector be resolved into *two* components only, one along Ox and the other in the perpendicular plane yOz, the latter components taken together are equivalent to the resolved part of the resultant in the plane yOz.

2·5. Vectors may be multiplied and divided by scalar numbers. Thus, if we take n equal vectors $\overline{AB}$ and compound them together, we get a vector $\overline{AC}$ such that $\overline{AC} = n\overline{AB}$; and conversely $\overline{AB} = \dfrac{1}{n}\overline{AC}$.

Note that relations of the form

$$\overline{AC} = n\overline{AB}, \quad \text{or} \quad p\overline{AB} + q\overline{AC} = 0,$$

imply that the points A, B, C are in the same straight line.

2·6. Centroids or Mean Centres. If m_1, m_2, m_3, ... m_n be a set of scalar magnitudes associated with a set of points A_1, A_2, A_3, ... A_n, the centroid or mean centre of the points for the given magnitudes is the point obtained by the following process: Divide A_1A_2 at B_1 so that

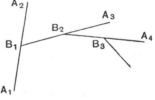

$$m_1 A_1 B_1 = m_2 B_1 A_2;$$

divide $B_1 A_3$ at B_2 so that

$$(m_1 + m_2) B_1 B_2 = m_3 B_2 A_3;$$

divide $B_2 A_4$ at B_3 so that $(m_1 + m_2 + m_3) B_2 B_3 = m_4 B_3 A_4$. Proceed in this way until all the points have been connected, then the last point of division B_{n-1}, usually denoted by the letter G, is called the *centroid* or *mean centre*.

2·61. In order to shew that this process leads in general to a unique point, i.e. that the point determined by the process is independent of the order in which the points A_1, A_2, ... A_n are joined, we shall first prove that

$$m_1 \overline{A_1 G} + m_2 \overline{A_2 G} + \dots + m_n \overline{A_n G} = 0 \quad \dots\dots (1).$$

Assume that this formula is true for the first r points, i.e. that

$$m_1 \overline{A_1 B_{r-1}} + m_2 \overline{A_2 B_{r-1}} + \dots + m_r \overline{A_r B_{r-1}} = 0.$$

Now the next step in the process is to divide $B_{r-1}A_{r+1}$ at B_r so that
$$(m_1 + m_2 + \ldots + m_r)\,\overline{B_{r-1}B_r} = m_{r+1}\,\overline{B_r A_{r+1}},$$
therefore, by adding the last two lines,
$$m_1\overline{A_1 B_r} + m_2\overline{A_2 B_r} + \ldots + m_r\overline{A_r B_r} + m_{r+1}\overline{A_{r+1}B_r} = 0.$$

It follows that if the formula (1) is true for the centroid of r points it is also true for the centroid of $r+1$ points; but it is true for two points, since, by hypothesis,
$$m_1\overline{A_1 B_1} + m_2\overline{A_2 B_1} = 0.$$
Therefore the formula (1) is true for the centroid of any number of points.

Now, if by taking the points in a different order we arrive at a centroid G', we can shew similarly that
$$m_1\overline{A_1 G'} + m_2\overline{A_2 G'} + \ldots + m_n\overline{A_n G'} = 0 \quad \ldots\ldots(2);$$
and by subtracting (1) from (2) we get
$$(m_1 + m_2 + \ldots + m_n)\,\overline{GG'} = 0.$$

Hence G' must coincide with G unless $m_1 + m_2 + \ldots + m_n = 0$. In the latter case there is no centroid at a finite distance, because the last step in the process of finding the centroid consists in dividing a line in the ratio $m_n : m_1 + m_2 + \ldots + m_{n-1}$, i.e. in the ratio $1 : -1$.

2·7. Centroid Method of Compounding Vectors. To shew, with the notation of **2·6**, that, if O be any other point, the resultant of n vectors $m_1\overline{OA_1}$, $m_2\overline{OA_2}$, ... $m_n\overline{OA_n}$ is $(m_1 + m_2 + \ldots + m_n)\,\overline{OG}$, where G is the centroid of the points $A_1, A_2, \ldots A_n$ for the magnitudes $m_1, m_2, \ldots m_n$.

This follows at once by substituting
$$\overline{OA_1} = \overline{OG} + \overline{GA_1}, \quad \overline{OA_2} = \overline{OG} + \overline{GA_2}, \text{ etc.},$$
so that
$$m_1\overline{OA_1} + m_2\overline{OA_2} + \ldots + m_n\overline{OA_n}$$
$$= (m_1 + m_2 + \ldots + m_n)\,\overline{OG} + (m_1\overline{GA_1} + m_2\overline{GA_2} + \ldots + m_n\overline{GA_n}),$$
and by **2·61** (1) the sum of the terms in the last bracket is zero, therefore
$$m_1\overline{OA_1} + m_2\overline{OA_2} + \ldots + m_n\overline{OA_n} = (m_1 + m_2 + \ldots + m_n)\,\overline{OG}.$$

2·71. When reference is made to the centroid of a set of points without mention of any associated magnitudes it is understood that the magnitudes are equal; thus the centroid of a triangle ABC is a point G such that

$$\overline{AG} + \overline{BG} + \overline{CG} = 0.$$

2·72. It may be noticed that if **P, Q, R** are vectors in the lines OA, OB, OC, then the resultant vector is

$$\left(\frac{P}{OA} + \frac{Q}{OB} + \frac{R}{OC}\right)\overline{OG},$$

where G is the centroid of the points A, B, C for the magnitudes P/OA, Q/OB, R/OC; for a vector **P** in the line OA is the same as $\frac{P}{OA}\overline{OA}$.

2·8. We began this chapter with a statement that certain physical quantities were to be classed together as vectors and then proceeded to define the properties of vectors and shew how they can be compounded. In order, therefore, to satisfy ourselves that a physical quantity such as force or acceleration is rightly described as a vector, we need adequate reasons for stating that

(i) it possesses direction,

(ii) it conforms to the laws $\overline{AB} = -\overline{BA}$,

and $\overline{AB} + \overline{BC} = \overline{AC}$.

EXAMPLES

1. ABC is a triangle. Prove that the magnitude of the resultant of vectors $\overline{AB}$, $2\overline{BC}$ and $3\overline{CA}$ is $(b^2 + c^2 + 2bc\cos A)^{\frac{1}{2}}$ and that its direction is that of the diagonal through A of the parallelogram of which AB, AC are adjacent sides.

2. $ABCDEF$ is a regular hexagon. Prove that

$$\overline{AB} + \overline{AC} + \overline{AD} + \overline{AE} + \overline{AF} = 3\overline{AD}.$$

3. AA', BB', CC', DD' are parallel edges of a parallelepiped, of which AC' is a diagonal. Prove that

$$\overline{AB} + \overline{AC} + \overline{AD} + \overline{AA'} + \overline{AB'} + \overline{AC'} + \overline{AD'} = 4\overline{AC'}.$$

4. Prove that, if G is the middle point of AB and G' is the middle point of $A'B'$, then $\overline{AA'} + \overline{BB'} = 2\overline{GG'}$.

5. Prove that, if G is the centroid of n points A_1, A_2, ... A_n, and G' is the centroid of n points B_1, B_2, ... B_n, then

$$\overline{A_1B_1} + \overline{A_2B_2} + ... + \overline{A_nB_n} = n\overline{GG'}.$$

6. Prove that, if A, B, C are any three points and G is a point such that $\overline{AG} + \overline{BG} + \overline{CG} = 0$, then G is the intersection of the medians of the triangle ABC.

7. Prove that, if A, B, C, D are any four points and G is a point such that $\overline{AG} + \overline{BG} + \overline{CG} + \overline{DG} = 0$, then G lies on the lines which join each point to the centroid of the other three.

8. Prove that the lines which join the middle points of the opposite edges of a tetrahedron meet in a point and bisect one another.

9. Prove that, if H is the orthocentre and O is the circumcentre of a triangle ABC, then

(i) $\overline{AH} + \overline{BH} + \overline{CH} = 2\overline{OA} + 2\overline{OB} + 2\overline{OC}$;

(ii) $\overline{AH} \tan A + \overline{BH} \tan B + \overline{CH} \tan C = 0$;

(iii) $\overline{AO} \sin 2A + \overline{BO} \sin 2B + \overline{CO} \sin 2C = 0$.

10. Shew that, if $m\overline{OA} + n\overline{OB} + p\overline{OC} = 0$ and $m + n + p = 0$, then the points A, B, C are collinear.

Chapter III

FORCES ACTING AT A POINT

3·1. In Chapter I we claimed that the idea of force is a primary conception attained by experience, and we inferred that a force possesses magnitude and direction and a definite line of action, and that its point of application may be assumed to be any point in the line of action, provided that the point is rigidly connected with the body upon which the force acts.

3·11. In order to proceed further in constructing a theory of forces we need some further knowledge of the properties of force, and it is open to us either to appeal to experiment and base our theory on so-called 'experimental law', or frankly to state that the system of Mechanics which we are about to develop needs a certain hypothesis or postulate concerning the nature of force, and that, taking for granted this hypothesis, the system of Mechanics which we erect upon this basis leads to results in conformity with everyday experience so far as it can be tested, and thence infer that our basic hypothesis is a sound one.

3·12. If we were to adopt the first alternative we should start from the 'law of moments' de-rived from experiments with the lever: i.e. the property of the lever known to Archimedes, that if AB is a straight lever which can turn about a fulcrum

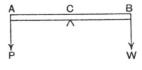

C, and weights P and W suspended from A and B keep the lever at rest in any position, then $P.AC = W.BC$.

It would be possible by a generalization of this result to build up a complete system of Statics.

3·13. On the ground of simplicity in developing the theory we prefer, however, to take the second alternative and adopt as our further postulate about force the statement that a force is a vector, so that the magnitude and direction of the resultant of two forces is determined by the vector law of addition; but inasmuch as a force is a vector *localized in a line* due care must be exercised in order to determine in what line the resultant force is localized. This determination is embodied in the proposition known as the Parallelogram of Forces, which states that: *If two forces acting at a point are represented in magnitude and direction by the sides of a paral-* 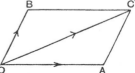 *lelogram drawn from that point, then their resultant is represented by the diagonal of the parallelogram drawn from that point.*

Thus if $OACB$ is a parallelogram and OA, OB represent forces acting at O, then OC represents the resultant of the forces OA, OB.

3·14. The enunciation of the foregoing proposition refers to forces 'acting at a point'. We have already stated that forces can only act on a body, so that the 'point' must be a point of a body, and there is no reason why the body should not be a single particle.

3·15. Consider two forces acting upon a body and represented by the lines AB, BC. In accordance with the vector law of addition, the resultant force is represented in magnitude and direction by the vector AC, but this is *not the position of the resultant.* B is the point of intersection of the forces and 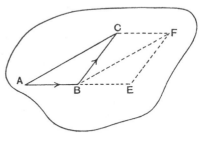 the only point on their lines of action at which *both* forces can be considered to act. Hence the resultant is represented by a line through B equal and parallel to AC; and this accords with the parallelogram law, for in order that the forces may be represented by lines drawn *from* a point we must produce AB to E so that $BE=AB$, then the forces AB, BC are equivalent to BE, BC and have a resultant represented by the diagonal BF of the parallelogram $BEFC$, which is easily seen to be equal and parallel to AC. It is necessary to warn the student against assuming that the vector law

$$\overline{AC} = \overline{AB} + \overline{BC}$$

gives anything more than the magnitude and direction of the resultant.

3·16. We propose then to take the *parallelogram of forces* as our fundamental hypothesis or postulate, and the evidence for the truth of the hypothesis is not to be found in a formal proof based upon some other assumption about force, but in the fact that the theory of Statics built up from this hypothesis accords with experience so far as it can be tested.

3·2. Triangle of Forces. When a body is in equilibrium under the action of three forces, it follows that if two of the forces intersect, the third force must pass through the point of intersection, because it must be equal and opposite to the resultant of the other two.

A simple deduction from the parallelogram of forces is the following theorem, known as the *Triangle of Forces: when three forces acting at a point can be represented in magnitude and direction by the sides of a triangle taken in order the forces are in equilibrium.*

Let the lines OA, OB, OC represent the three forces P, Q, R acting at O. Complete the parallelogram $OADB$. Then AD being equal and parallel to OB represents the force Q in magnitude and direction, and if the forces P, Q, R can be represented by the sides of a triangle taken in order, then OAD

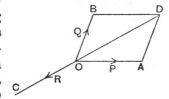

is such a triangle because OA, AD represent P and Q. Therefore DO represents R. But OD is the resultant of P and Q. Therefore R is equal and opposite to the resultant of P and Q and the three forces are in equilibrium.

Conversely, *if three forces acting at a point are in equilibrium and a triangle be drawn with its sides parallel to the directions of the forces, then the lengths of the sides will be proportional to the magnitudes of the corresponding forces.* For with the same figure and construction, since the forces are in equilibrium, therefore R balances the resultant OD of P and Q; therefore DO represents R, while OA and AD represent P and Q. Hence OAD is a triangle having its sides OA, AD, DO parallel to the forces and also proportional to their magnitudes.

3·21. Lami's Theorem. *If three forces acting at a point are in equilibrium, then each force is proportional to the sine of the angle between the other two.*

For, making use of the result and figure of the preceding theorem, if the forces P, Q, R acting at O are in equilibrium, we have
$$P : Q : R = OA : AD : DO$$
$$= \sin ODA : \sin DOA : \sin OAD$$
$$= \sin BOC : \sin COA : \sin AOB,$$

using supplementary angles.

3·211. Example. *The ends of a light string $ABCD$ are fixed. Weights are fastened to the string at B and C, and the parts AB, BC, CD are inclined to the horizontal at angles of $60°$, $30°$ and $55°$. The weight at B is 22 lb., determine that at C.*

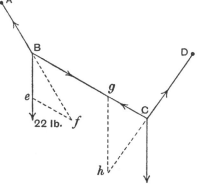

This problem can be solved graphically by drawing triangles of forces or directly by calculation.

Thus, at B there are three forces in equilibrium, namely the weight of 22 lb. vertically downwards, which we represent on a convenient scale by a length Be, and the tensions of the strings BA, BC. From e we draw a parallel to BC to meet AB produced in f and thus get a triangle Bef with its sides parallel to the forces. Then by the converse of the triangle of forces ef, fB represent the tensions in BC and BA and by measurement they represent 22 lb. and 38 lb.

Next, at the point C there are three forces in equilibrium, namely a weight acting vertically downwards and the tensions of two strings CB, CD whose directions are known, and we have found that the tension in CB is 22 lb. So we construct a triangle of forces at C, by marking off on CB a length $Cg = fe$ and drawing a vertical line gh to meet DC

produced in h. Then the sides of the triangle Cgh are in the directions
of the three forces at C and we find by measurement that $gh = 38$ and
$hC = 33$. Hence, to this degree of accuracy the weight at C is 38 lb. and
the tension in CD is 33 lb.

Alternatively, we may use Lami's Theorem. Thus, if we denote the
tensions in AB, BC, CD by T_1, T_2, T_3, we have at B

$$\frac{T_1}{\sin 60°} = \frac{T_2}{\sin 150°} = \frac{22}{\sin 150°},$$

whence we get $T_2 = 22$ lb. and $T_1 = 38 \cdot 1$ lb.

Again, if W denotes the weight at C,

$$\frac{W}{\sin 95°} = \frac{T_3}{\sin 120°} = \frac{T_2}{\sin 145°},$$

giving $W = 38 \cdot 2$ lb. and $T_3 = 33 \cdot 2$ lb.

3·22. Polygon of Forces. *If any number of forces acting at
a point can be represented in magnitude and direction by the sides
of a polygon taken in order, then the forces are in equilibrium.*

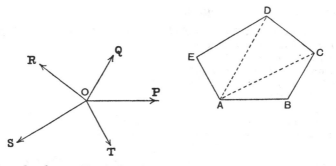

Let the forces **P**, **Q**, **R**, **S**, **T** acting at O be represented in
magnitude and direction by the sides AB, BC, CD, DE, EA
of the polygon $ABCDE$, then the forces are in equilibrium.

For if we compound the forces by the vector law, step by
step, we have

$$\mathbf{P} + \mathbf{Q} = \overline{AB} + \overline{BC} = \overline{AC}$$
$$\mathbf{P} + \mathbf{Q} + \mathbf{R} = \overline{AC} + \overline{CD} = \overline{AD}$$
$$\mathbf{P} + \mathbf{Q} + \mathbf{R} + \mathbf{S} = \overline{AD} + \overline{DE} = \overline{AE}$$
$$\mathbf{P} + \mathbf{Q} + \mathbf{R} + \mathbf{S} + \mathbf{T} = \overline{AE} + \overline{EA} = 0,$$

where in each equation the resultant, on the right, of the forces
named on the left acts at the point O. It follows that the forces
are in equilibrium.

We have set out the composition of the forces step by step but we might have said at once that vectors AB, BC, CD, DE, EA have a zero resultant, and hence the given forces are in equilibrium.

3·23. It should be noted that in **3·2, 3·22** it is essential that the forces should *act at a point*. It will be seen later that the theorems **3·2, 3·22** are *not true* if the forces are represented *in position* as well as in magnitude and direction by the sides of the triangle or polygon.

There is no converse proposition to the polygon of forces—if there were, then by analogy from the converse of the triangle of forces it would be that if a number of forces acting at a point were in equilibrium and a polygon were drawn with its sides parallel to the forces the sides of the polygon would be proportional to the magnitudes of the corresponding forces; but this is clearly not true because equiangular polygons are not necessarily similar.

3·24. It should also be noted that in the polygon of forces the forces need not necessarily all be in the same plane.

3·3. Composition of Forces acting at a point. There are three methods of finding the resultant of a number of forces acting at a point and they are the same as the methods given in Chapter II for finding the resultant of a number of vectors.

(i) *The Graphical Method.*

Let P, Q, R, S be the forces acting at O. Starting from any point A, draw lines AB, BC, CD, DE to represent P, Q, R, S

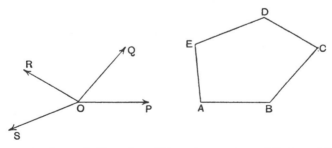

in magnitude and direction. Then by vector addition AE represents the resultant of P, Q, R, S, and a force equal and parallel to AE acting at O is the required resultant.

(ii) *The Analytical Method* as in **2·4**, resolving the forces along rectangular axes, adding up the components along each axis and then compounding the results.

(iii) *The Centroid Method*, which may be enunciated as follows: *If any number of forces acting at a point O are represented by the vectors $m_1\overline{OA}_1$, $m_2\overline{OA}_2$, $m_3\overline{OA}_3$, ..., then their resultant is represented by $(m_1+m_2+m_3+...)\,\overline{OG}$, where G is the centroid of the points A_1, A_2, A_3, ... for the multiples m_1, m_2, m_3,....*

The proof of this theorem is contained in 2·6 and 2·7.

We observe that it is not necessary that the points O, A_1, A_2, A_3, ... should all lie in the same plane.

3·31. Expressions for the Resultant. *Let P_1, P_2, P_3, ... be any number of forces acting at a point and let θ_{rs} denote the. angle between the forces P_r and P_s, then the resultant force R is given in magnitude by*

$$R^2 = \Sigma P_r{}^2 + 2\Sigma P_r P_s \cos\theta_{rs},$$

where the latter sum includes the products of all the forces taken in pairs.

(i) When the forces are coplanar: let α_1, α_2, α_3, ... be the angles which the forces P_1, P_2, P_3, ... make with a fixed direction Ox. Then as in 2·4 (i)

$$R^2 = \{\Sigma\,(P_r\cos\alpha_r)\}^2 + \{\Sigma\,(P_r\sin\alpha_r)\}^2$$
$$= \Sigma\{P_r{}^2(\cos^2\alpha_r+\sin^2\alpha_r)\} \qquad\qquad$$
$$\qquad\qquad + 2\Sigma\{P_rP_s(\cos\alpha_r\cos\alpha_s+\sin\alpha_r\sin\alpha_s)\}$$
$$= \Sigma P_r{}^2 + 2\Sigma P_r P_s\cos(\alpha_r-\alpha_s)$$
$$= \Sigma P_r{}^2 + 2\Sigma P_r P_s\cos\theta_{rs}.$$

(ii) When the forces are not all in the same plane, as in 2·4 (ii), let a typical force P_r make angles α_r, β_r, γ_r with three axes Ox, Oy, Oz mutually at right angles. Then

$$R^2 = \{\Sigma\,(P_r\cos\alpha_r)\}^2 + \{\Sigma\,(P_r\cos\beta_r)\}^2 + \{\Sigma\,(P_r\cos\gamma_r)\}^2$$
$$= \Sigma\{P_r{}^2(\cos^2\alpha_r+\cos^2\beta_r+\cos^2\gamma_r)\}$$
$$+ 2\Sigma\{P_rP_s(\cos\alpha_r\cos\alpha_s+\cos\beta_r\cos\beta_s+\cos\gamma_r\cos\gamma_s)\}.$$

Then as in 2·4 (ii)

$$\cos^2\alpha_r+\cos^2\beta_r+\cos^2\gamma_r = 1,$$

and it can easily be proved* that

$$\cos \alpha_r \cos \alpha_s + \cos \beta_r \cos \beta_s + \cos \gamma_r \cos \gamma_s = \cos \theta_{rs}.$$

Therefore $\qquad R^2 = \Sigma P_r{}^2 + 2\Sigma P_r P_s \cos \theta_{rs}.$

3·4. Conditions of Equilibrium of Forces acting at a point. Forces acting at a point are in equilibrium when their resultant is zero.

(i) *Graphically*—if the forces are represented in magnitude and direction by lines forming the successive sides of a polygon, the polygon must be closed.

(ii) *Analytically*—if the forces are resolved along rect-angular axes, then for forces in one plane as in **2·4** (i)

$$R^2 = X^2 + Y^2,$$

and R can only be zero if X and Y are both zero; and for forces not all in one plane as in **2·4** (ii)

$$R^2 = X^2 + Y^2 + Z^2,$$

and R can only be zero if X, Y and Z are all zero. Hence a necessary condition of equilibrium is that the sum of the resolved parts of the forces in any direction is zero—*any* direc-tion because the choice of direction of the axes in relation to the forces is arbitrary.

(iii) *Centroid Method*—forces acting at a point O represented by vectors $m_1 \overline{OA_1}$, $m_2 \overline{OA_2}$, $m_3 \overline{OA_3}$, ... are clearly in equili-brium if and only if O is the centroid of the points $A_1, A_2, A_3, ...$ for the multiples $m_1, m_2, m_3, ...$

---

* Let OP, OQ be the directions of the forces P_r, P_s. Take OP of unit length and draw PN perpendicular to OQ, PM per-pendicular to the plane xOy and ML per-pendicular to Ox. Then OL, LM, MP being equal to the projections of OP on the axes Ox, Oy, Oz are of lengths $\cos \alpha_r$, $\cos \beta_r$, $\cos \gamma_r$; and θ_{rs} is the angle POQ, therefore

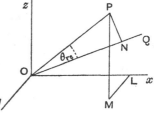

$$\cos \theta_{rs} = ON = \text{projection of } OP \text{ on } OQ$$
$$= \text{sum of projections on } OQ \text{ of}$$
$$OL, LM, MP$$
$$= OL \cos \alpha_s + LM \cos \beta_s + MP \cos \gamma_s$$
$$= \cos \alpha_r \cos \alpha_s + \cos \beta_r \cos \beta_s$$
$$\quad + \cos \gamma_r \cos \gamma_s.$$

3·41. In many examples we are only concerned with three coplanar forces. If the forces are not all parallel, two of them must meet in a point and for equilibrium it is necessary that the third force shall balance the resultant of the other two, but this resultant passes through the point of intersection of the first two, therefore the third force must pass through the point of intersection of the first two. Hence if three forces are not all parallel they can only be in equilibrium when they meet in a point. The further condition necessary for equilibrium being that the magnitudes of the forces must be proportional to the lengths of the sides of any triangle drawn parallel to them (**3·2**), or each force proportional to the sine of the angle between the other two (**3·21**). The case in which the forces are parallel will be considered later.

3·42. Oblique Resolution. A force may be resolved into two components in any assigned directions which are coplanar with it. Thus if OC represents a force R and Ox, Oy are any lines making angles α, β with OC and in the same plane, by drawing CA, CB parallel to Oy, Ox we form a parallelogram and the force R or OC is equivalent to components P, Q represented by OA, OB such that

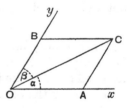

$$P : Q : R = OA : AC : OC$$
$$= \sin\beta : \sin\alpha : \sin(\alpha+\beta);$$

or
$$P = \frac{\sin\beta}{\sin(\alpha+\beta)}R \text{ and } Q = \frac{\sin\alpha}{\sin(\alpha+\beta)}R.$$

3·43. The centroid method may also sometimes be used conveniently for resolving a force into components along assigned directions. Thus a force R acting along a line OG may be resolved into three forces P_1, P_2, P_3 along lines OA_1, OA_2, OA_3 if we can find points A_1, A_2, A_3 on these lines and multiples m_1, m_2, m_3 such that G is the centroid of A_1, A_2, A_3 for these multiples; for in such a case

$$\frac{P_1}{m_1 OA_1} = \frac{P_2}{m_2 OA_2} = \frac{P_3}{m_3 OA_3} = \frac{R}{(m_1+m_2+m_3)OG}.$$

3·431. Example. It is easy to prove that the orthocentre H of a triangle ABC is the centroid of the points A, B, C for multiples $\tan A$, $\tan B$, $\tan C$. Hence if O be any other point, a force R along OH is equivalent to forces P_1, P_2, P_3 along OA, OB, OC such that

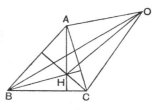

$$\frac{P_1}{OA\tan A} = \frac{P_2}{OB\tan B} = \frac{P_3}{OC\tan C}$$

$$= \frac{R}{(\tan A + \tan B + \tan C)\,OH} = \frac{R}{OH\tan A\tan B\tan C}.$$

The point O need not be in the plane ABC.

3·5. Examples. (i) *Two forces P, Q act at a point along two straight lines making an angle α with each other and R is their resultant: two other forces P', Q' acting along the same two lines have a resultant R'. Prove that the angle between the lines of action of the resultants is*

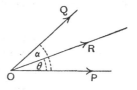

$$\cos^{-1}\{PP' + QQ' + \cos\alpha\,(PQ' + P'Q)\}/RR'.$$
[S.]

Let the resultants R, R' make angles θ, θ' with the line of action of P and P'. By resolving along and perpendicular to this line, we get

$$R\cos\theta = P + Q\cos\alpha, \quad R\sin\theta = Q\sin\alpha,$$

and
$$R'\cos\theta' = P' + Q'\cos\alpha, \quad R'\sin\theta' = Q'\sin\alpha.$$

Therefore

$$R R'\cos(\theta - \theta') = R R'(\cos\theta\cos\theta' + \sin\theta\sin\theta')$$
$$= (P + Q\cos\alpha)(P' + Q'\cos\alpha) + QQ'\sin^2\alpha$$
$$= PP' + QQ' + \cos\alpha\,(PQ' + P'Q),$$

whence the result follows.

(ii) *Two forces given in magnitude act each through a fixed point, and are inclined at a constant angle θ; shew that their resultant also passes through a fixed point A.*

If θ varies, shew that the locus of A is a circle. [S.]

Let B, C be the given fixed points, and P, Q the magnitudes of the given forces. On BC construct a segment of a circle containing an angle $BOC = \theta$. Then forces P, Q acting along OB, OC will have a resultant along a line OA which cuts the circle again in a point A.

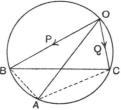

Then so long as the angle BOC between the forces is constant, since the forces are of given magnitudes the angle which their resultant makes with either force is also constant, i.e. the angle BOA is constant and therefore A is a fixed point.

Again, when the angle θ varies, we still have

$$\frac{BA}{AC} = \frac{\sin BOA}{\sin AOC} = \frac{Q}{P} \quad (3\cdot42);$$

but B and C are fixed points, therefore the locus of A is a circle. (Circle of Apollonius.)

(iii) $ABCD...$ *is a polygon of* n *sides, and forces act at a point parallel and proportional to* AB, $2BC$, $3CD$, *etc. Shew that their resultant is parallel and proportional to*

$$(n-1)\, OA,$$

where O is the centroid of all the points B,
C, D, ... excluding A.   [S.]

Let LA be the nth side of the polygon; then, as regards magnitude and direction,

since $\qquad \overline{AB} = \overline{AO} + \overline{OB},$

$\qquad\qquad\quad \overline{BC} = \overline{BO} + \overline{OC},$

and so on, therefore we have

$\overline{AB} + 2\overline{BC} + 3\overline{CD} + ... + n\overline{LA}$
$= \overline{AO} + \overline{OB} + 2\overline{BO} + 2\overline{OC} + 3\overline{CO} + 3\overline{OD} + ... + n\overline{LO} + n\overline{OA}$
$= (n-1)\,\overline{OA} + \overline{BO} + \overline{CO} + \overline{DO} + ... + \overline{LO}.$

But since O is the centroid of the points B, C, D, ... L, therefore

$$\overline{BO} + \overline{CO} + \overline{DO} + ... + \overline{LO} = 0 \quad (3\cdot4 \text{ (iii))},$$

and the required resultant is represented by $(n-1)\,\overline{OA}$.

3·6. Systems of Particles. Internal and External Forces.

So far we have only considered the case of forces acting 'at a point' or on a single particle. It is a simple step further to the case of two or more particles in equilibrium under the action of given forces and the interactions of the particles upon one another. For example, let A, B, C be three particles connected by light threads which are kept taut by applying

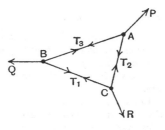

forces P, Q, R to the particles A, B, C respectively. The interaction between each pair of particles is then a tension in the string joining them, which acts in opposite directions on the two particles. If we denote the tensions in BC, CA, AB by

T_1, T_2, T_3, then, considering each particle separately,

A is in equilibrium under the forces P, T_2, T_3,

B is in equilibrium under the forces Q, T_3, T_1,

and C is in equilibrium under the forces R, T_1, T_2.

And the system of three particles A, B, C is in equilibrium under the action of the three forces P, Q, R, because when we add together the nine forces previously named they include three pairs of equal and opposite forces which cancel out.

This example serves to illustrate the difference between 'internal' and 'external' forces. Considering the three particles *as one system* the forces P, Q, R are 'externally applied forces', i.e. forces applied by some external agent, and the tensions T_1, T_2, T_3 are 'internal forces' since they constitute the mutual reactions of the particles upon one another. But if, on the other hand, we consider a single particle, say A, as a self-contained system, then we do not classify some of the forces P, T_2, T_3 as internal, they are all regarded in the same way as externally applied forces.

There is no hard and fast rule that such and such a force should always be regarded as either internal or external. Thus the weight of this book, or the force with which the Earth attracts it, is an external force when we are considering the equilibrium of the book as a separate entity; but when we consider the Earth and all things on the Earth as an entity attracted by the sun and moon, then the weight of the book and the counterbalancing reaction of the body which supports it are internal forces which cancel out.

3·7. Equilibrium under constraint. Smooth and Rough Bodies.

The freedom of a body is often restricted by passive obstacles, e.g. a donkey tethered to a post, a bead threaded on a wire; or by conditions imposed in mathematical language, e.g. a particle shall remain on a given material curve or surface, the ends of a rod shall lie upon a given curve and so forth. In such cases the conditions imply the existence of a force or forces of constraint, such as the tension of the tethering rope, the pressure of the wire on the bead, of the curve or surface on the particle, and of the curve on the ends of the rod. Such forces are *passive* in the sense that no more force will be called into play than is necessary to maintain equilibrium.

When two bodies are in contact at a point, the mutual reaction R between them is not in general at right angles to their common tangent plane at the point of contact but inclined to the normal at an angle α say. This force R can then be resolved into components $R \cos \alpha$ along the common normal and $R \sin \alpha$ in the tangent plane. The former component is called the *normal reaction* and the

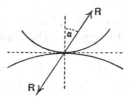

component that lies in the tangent plane is called **Friction**; the latter is a measure of the resistance which the bodies themselves offer to any attempt to make the one slide over the other. Friction depends upon the nature of the substances in contact and will be discussed fully in a later chapter. Bodies which are incapable of exerting any frictional force or between which the force of friction is so small as to be negligible in comparison with the other forces in action are said to be *smooth*. In every case of contact of smooth bodies the mutual reaction between them must act along the common normal at the point of contact.

When a rod passes through a smooth ring the mutual reaction between them is at right angles to the rod, and the ring can only be in equilibrium if the other forces acting upon it when resolved along the rod balance one another, for otherwise the ring would slide along the rod.

When a particle is to remain in contact with a smooth material surface the reaction of the surface on the particle can only be along the normal in one direction and it is necessary that the external forces acting on the particle shall press it against the surface and have a resultant in no other direction than the normal, for otherwise the particle would slide on the surface.

3·71. Examples. (i) *A bead of weight W can slide on a smooth circular wire in a vertical plane. The bead is attached by a light thread to the highest point of the wire, and in equilibrium the thread is taut and makes an angle* *θ with the vertical. Find the tension of the thread and the reaction of the wire on the bead.*

Let B be the bead, AB the thread, AOC the vertical diameter of the circle, and O the centre. Then the angle

$$OBA = OAB = \theta, \quad \text{and} \quad BOC = 2\theta.$$

Hence, if T denotes the tension and R the reaction, by Lami's Theorem

$$T : R : W = \sin(R, W) : \sin(W, T) : \sin(T, R)$$
$$= \sin 2\theta : \sin \theta : \sin \theta,$$

therefore $T = 2W \cos \theta$ and $R = W$.

Alternatively, BAO is a triangle whose sides are parallel to the forces, so that

$$T : R : W = BA : OB : AO,$$

whence $\qquad T = W \dfrac{BA}{AO} = 2W \cos \theta, \quad \text{and} \quad R = W.$

(ii) *The ends of a light string are attached to two smooth rings of weights w, w' and the string carries a third smooth ring of weight W which can slide upon it; the rings w, w' are free to slide on two fixed rods inclined at angles α and β to the vertical. Prove that, if ϕ be the angle which either part of the string makes with the vertical, then in equilibrium*

$$\cot \phi : \tan \beta : \tan \alpha = W : W + 2w' : W + 2w.$$

Let P, Q, R be the rings of weights W, w, w' respectively. It will appear in a later chapter that the tension of a string is not affected by

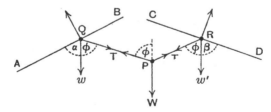

passing through a *smooth* ring, so that the tensions in the parts PQ, PR of the string are of equal magnitude T say. Then since two of the forces acting on the ring P are equal they must be equally inclined to the third force, namely the weight, therefore PQ, PR are equally inclined to the vertical at an angle ϕ say.

By resolving vertically for the ring P, we get

$$2T \cos \phi = W \qquad \dots\dots\dots\dots\dots\dots\dots(1).$$

Consider next the ring Q. It is in equilibrium under the action of its weight w, the tension T of the string QP and the reaction of the smooth rod AB on which the ring can slide. This reaction is at right angles to the rod and since it balances w and T its line of action (produced backwards) must fall between w and T. This fact tells us how to place the rod AB in the diagram and by a like argument we find the position of the rod CD.

Then, by resolving along BA for the ring Q, we have

$$w \cos \alpha + T \cos (\alpha + \phi) = 0 \ldots\ldots\ldots\ldots\ldots\ldots(2).$$

By eliminating T between (1) and (2), we get

$$2w \cos \alpha \cos \phi + W \cos (\alpha + \phi) = 0,$$

or $$(2w + W) \cot \phi = W \tan \alpha.$$

Similarly $$(2w' + W) \cot \phi = W \tan \beta,$$

and these are the required results.

It will be observed that we might easily find the reactions of the rods on the rings Q and R by using Lami's Theorem; but it is not necessary to introduce them into our equations for the purpose of the problem, and we avoid doing so by resolving along the rod. Of course a partial application of Lami's Theorem to get the ratio of T to w gives the same result.

EXAMPLES

1. Find graphically and by calculation the resultant of two forces of 5 lb. and 3 lb. acting at an angle of 100°, and determine the angle which the resultant makes with the larger force.

2. Forces of 7 lb., 10 lb. and 12 lb. acting at a point are in equilibrium; determine graphically and by calculation the angles between their lines of action.

3. AB is a light string 24 in. long; its upper end A is fastened to a fixed point, and B is attached to a weight of 10 lb. Determine the force required to hold B at a distance of 12 in. from the vertical through A, when the force is applied (i) horizontally, (ii) at right angles to AB.

4. A square $ABCD$, of side 4 ft., is acted on by a force of 8 lb. along the diagonal AC and a force of 4 lb. along the diagonal DB. These two forces can be balanced by a third force acting through a point P in BC. Find graphically or otherwise the length of BP.

5. A force of 12 lb. is resolved into two components, one of which is 5 lb. in a direction making an angle of 30° with the former force. Find the other component in magnitude and direction.

6. Three ropes, all in the same vertical plane, meet at a point, and there support a block of stone. They are inclined to the horizontal at angles 35°, 100° and 160°. The tensions in the first two ropes are 200 lb. and 150 lb. Find graphically the tension in the third rope and the weight of the block of stone, and verify by calculation.

7. A string $ABCD$ hangs from fixed points A, D, carrying a weight of 12 lb. at B and a weight W at C. AB is inclined at 60° to the horizontal, BC is horizontal and CD is inclined at 30° to the horizontal. By drawing triangles of forces, or otherwise, find W.

8. Two forces act at a point and are such that if the direction of one is reversed the direction of the resultant is turned through a right angle. Prove that the two forces must be equal in magnitude.

9. Forces of $2, \sqrt{3}, 5, \sqrt{3}, 2$ lb. respectively, act at one of the angular points of a regular hexagon towards the five others in order. Find the magnitude and direction of the resultant. [S.]

10. Equal forces P act at a point parallel to the sides of a triangle ABC taken in order the same way round. Prove that the resultant R is given by $\qquad R^2 = P^2(3 - 2\cos A - 2\cos B - 2\cos C)$.

11. A weight of 10 lb. is supported by two strings which make angles of 30° and 60° with the vertical. Find the tensions in the strings.

12. A bead free to slide on a smooth circular wire in a vertical plane is attached by a fine taut thread to a given point in the vertical line through the centre of the circle. Shew that the pressure of the wire on the bead is independent of the length of the string.

13. Two beads of weights w and w' can slide on a smooth circular wire in a vertical plane. They are connected by a light string which subtends an angle 2β at the centre of the circle when the beads are in equilibrium on the upper half of the wire. Prove that the inclination α of the string to the horizontal is given by

$$\tan \alpha = \frac{w \sim w'}{w + w'} \tan \beta.$$

14. Weights w and w' are fastened to the ends of a light rod AB which is suspended from a point O by strings OA, OB and the vertical through O cuts AB in C. By considering triangles of forces, prove that

$$w.AC = w'.BC,$$

and that the tensions in OA, OB are in the ratio

$$AO.BC : BO.AC.$$

15. Three given weights P, Q, R, any two of which are together greater than the third, are attached to the ends of three strings the other ends of which are knotted together at a point O. The strings that carry the weights P, Q pass over two smooth pegs A and B and the weight R hangs between the pegs. Give a geometrical construction for finding the position of the point O.

16. A weight is supported by a light string passing over a smooth pulley and gently lowered on to a smooth inclined plane. Shew that if the string is slowly paid out so that the weight slides down the plane the pressure on the plane increases and the tension of the string decreases.

17. A small ring is capable of motion along a wire of circular shape, and is attracted by forces varying as the distance, and of equal absolute intensities, to two given external points in the plane of the circle. Shew that in any position of the ring the resultant attraction passes through a fixed point; and give a geometrical construction for a position of equilibrium. [S.]

18. A weight of 10 lb. hangs by a string from a fixed point. The string is drawn out of the vertical by applying a force of 5 lb. to the weight. In what direction must this force be applied in order that in equilibrium the deflection of the string from the vertical may have its greatest value? What is the amount of the greatest deflection?

19. O is any point in the plane of a triangle ABC; D, E, F are the middle points of the sides. Prove that the resultant of forces represented in magnitude and direction by OE, OF, DO is represented by OA.   [S.]

20. P is any point in the plane of a triangle ABC, and D, E, F are the middle points of its sides; prove that forces AP, BP, CP, PD, PE, PF are in equilibrium.   [S.]

21. A system of n forces acting at a point is represented in magnitude and direction by the lines A_1B_1, A_2B_2, ... A_nB_n, where A_1, A_2, ... A_n and B_1, B_2, ... B_n are the corners of two regular polygons. Find a line representing the resultant force in magnitude and direction.   [S.]

22. Any two points E, F are taken on the sides AB, CD of the parallelogram $ABCD$; G, H are the middle points of AC, EF respectively; prove that the resultant of forces acting at a point represented by EC, ED, FA, FB is represented in magnitude and direction by $4HG$.   [C.]

23. Three equal forces are represented in magnitude and direction by OA, OB, OC, where O is the circumcentre of the triangle ABC. Prove that the resultant is represented in magnitude and direction by the line joining O to the orthocentre of the triangle.

24. ABC is a triangle; prove that six forces represented by AH, BH, CH, AA', BB', CC' will be in equilibrium; where H is the orthocentre and AA', BB', CC' are diameters of the circumcircle.   [S.]

25. Forces P, Q, R acting at a point O are in equilibrium and a straight line meets their lines of action in A, B, C respectively; shew that, with certain conventions of sign,

$$\frac{P}{OA} + \frac{Q}{OB} + \frac{R}{OC} = 0.$$ [S.]

26. Four straight lines in a plane intersect, two at a time, in the six points A, B, C, D, E, F, and O is any other point. Find a geometrical construction for the resultant of the forces represented by OA, OB, OC, OD, EO, FO.   [S.]

27. Three forces acting at a point are parallel to the sides of a triangle ABC, taken in order, and proportional to the cosines of the opposite angles; shew that their resultant is proportional to

$$(1 - 8\cos A \cos B \cos C)^{\frac{1}{2}}.$$ [S.]

28. P is a point in the plane of a triangle ABC, forces act at P towards the angular points represented by $PA.\sin A$, $PB.\sin B$ and

$PC . \sin C$ respectively: shew that the resultant is

$$4PI . \cos\frac{A}{2} \cos\frac{B}{2} \cos\frac{C}{2},$$

where I is the incentre of the triangle.   [S.]

29. A particle P is attracted towards each of four points A, B, C, D by forces equal to $\mu_1 PA$, $\mu_2 PB$, $\mu_3 PC$, $\mu_4 PD$. Shew that it will rest in equilibrium only at the centroid of A, B, C, D, for multiples μ_1, μ_2, μ_3, μ_4.

Shew also that if a force Q be applied to the particle its position of equilibrium will be displaced from this position a distance

$$Q/(\mu_1 + \mu_2 + \mu_3 + \mu_4)$$

in the direction of Q.   [S.]

30. If O, A, B, C, D, ... Z be any fixed points, and if any points P, Q, R, ... be taken in AB, BC, CD, ... YZ, ZA respectively, so that

$$\frac{AP}{PB} = \frac{BQ}{QC} = \frac{CR}{RD} = ...,$$

shew that the resultant of the forces represented by OP, OQ, OR, ... is constant in magnitude and direction.   [I.]

31. Forces act at a point P, along the lines joining P to the vertices of a triangle ABC, and the magnitudes of the forces are

$(\sin 2B + \sin 2C) AP$, $(\sin 2C + \sin 2A) BP$, $(\sin 2A + \sin 2B) CP$

respectively; prove that their resultant passes through the centre of the 'nine points' circle of the triangle ABC, and find its magnitude.
                    [S.]

32. $ABCD$ is a quadrilateral inscribed in a circle whose centre is O. Prove that forces along AO, OB, CO, OD proportional to the areas BCD, CDA, DAB, ABC are in equilibrium.

33. I is the centre of the inscribed circle of a triangle ABC and D is any point. Prove that a force ID is equivalent to three forces $\frac{a}{2s} AD$, $\frac{b}{2s} BD$, $\frac{c}{2s} CD$, where $2s = a + b + c$.

34. Prove that, if E, F are the feet of the perpendiculars from two corners of a triangle ABC upon the opposite sides, a force P acting along EF can be replaced by forces $P\cos A$, $P\cos B$, $P\cos C$ acting along the sides.   [S.]

35. Forces are represented by the lines joining one corner of a parallelepiped to the middle points of the edges which do not meet in it. Prove that their resultant is represented by $5\frac{1}{2}$ times the diagonal through that corner.   [S.]

36. Two forces are represented by radii vectores drawn from the focus of an ellipse to the curve, the sum of the two being given. Shew that the locus of the extremity of the resultant is a straight line parallel to the minor axis.   [S.]

37. If $ABCD$ be a plane quadrilateral such that the triangles ABC, ADC are equal in area, O any point in the plane, H, K the middle points of AC, BD, and a, b, c, d the middle points of the sides, then forces $\frac{1}{2}Oa$, $\frac{1}{2}Ob$, $\frac{1}{2}Oc$, $\frac{1}{2}Od$ and OH are equivalent to forces OA, OB, OC, OD and KO. [S.]

38. The line of action of a force P cuts the sides of a triangle in given points. Shew how to find three forces acting along the sides of the triangle which shall have P for their resultant. [S.]

39. Forces P, Q, R, S, T acting at a point form a system in equilibrium. If the angles between (R, S), (S, T), (T, P) be α, β, γ respectively, find, by the polygon of forces or otherwise, what two forces, in the directions of P and R respectively, will balance Q. [S.]

40. O, O' are inverse points with respect to a circle centre C and P is a particle anywhere on the circle acted upon by forces $\dfrac{\mu'}{O'P^2}$ along PO' and $\dfrac{\mu}{OP^2}$ along OP. Shew that the resultant force on the particle is normal to the circle, provided

$$\frac{\mu'}{\mu} = \sqrt{\frac{CO'}{CO}}. \qquad [\text{S.}]$$

41. Three small particles are placed in a narrow, smooth, circular tube and repel each other in such a way that the force between any two of them is proportional to the product of their masses and their distance apart. Shew that when they are in equilibrium their masses are proportional to the sides of the triangle formed by drawing tangents to the tube at the positions of the particles. [S.]

ANSWERS

1. 5·36 lb., 33° 24′.   2. 92° 5′, 123° 37′, 144° 19′.
3. (i) $\frac{10}{3}\sqrt{3}$ lb., (ii) 5 lb.   4. $5\frac{1}{3}$ ft.
5. 8·06 lb. inclined at 18° 4′ to the resultant.   6. 146·6 lb., 312·6 lb.
7. 4 lb.   9. 10 lb. along the diagonal.   11. $5\sqrt{3}$ lb., 5 lb.
18. At right angles to the string; 30°.
21. nGG', where G, G' are the centres of the polygons.
31. $8PN \sin A \sin B \sin C$, where N is the 'nine points centre'.
39. $\{P\sin(\alpha+\beta+\gamma) + T\sin(\alpha+\beta) + S\sin\alpha\}/\sin(\alpha+\beta+\gamma)$
and $\{R\sin(\alpha+\beta+\gamma) + S\sin(\beta+\gamma) + T\sin\gamma\}/\sin(\alpha+\beta+\gamma)$.

Chapter IV

MOMENTS. PARALLEL FORCES. COUPLES

4·1. The Moment of a Force about a Point is defined to be the product of the force and the perpendicular distance of its line of action from the point.

Thus if P be a force and p the distance of its line of action from a point O, then the moment of P about O is Pp.

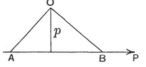

Also if the force P is represented by AB, then $Pp = AB.p =$ twice the area OAB.

This gives a geometrical representation of the moment of a force by an area.

The moment of a force about a point may be regarded as a measure of the tendency of the force to cause rotation about the point.

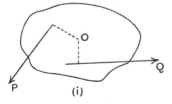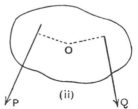

Thus if P and Q are forces applied to a disc in its plane and O is a point about which the disc can turn, in fig. (i) P and Q tend to cause rotation about O in the same (counter-clockwise) sense, and we should regard their moments as of the same sign; but in fig. (ii) P and Q tend to cause rotation about O in opposite senses and we should regard their moments as of opposite signs. Either sense—clockwise or counter-clockwise—may be chosen as the positive sense.

4·2. *When any number of coplanar forces act at a point the algebraical sum of their moments about any point in their plane is equal to the moment of their resultant about the same point.*

Let any number of forces $P_1, P_2, P_3, \ldots$ act at a point A, and let R be their resultant, and O the point about which moments are to be taken. Let the forces $R, P_1, P_2, P_3, \ldots$ be at the distances $r, p_1, p_2, p_3, \ldots$ from O and make angles $\theta, \alpha_1, \alpha_2, \alpha_3, \ldots$ with AO.

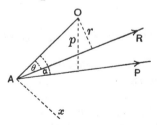

For simplicity only one of the forces $P_1, P_2, P_3, \ldots$ is shewn in the figure.

The algebraical sum of the moments about O of $P_1, P_2, P_3, \ldots$

$= \Sigma Pp = \Sigma P . OA \sin \alpha = OA . \Sigma P \sin \alpha$

$= OA \times$ alg. sum of resolved parts of the forces along a line Ax at right angles to OA

$= OA \times$ resolved part of R along Ax

$= OA \times R \sin \theta = R \times OA \sin \theta = Rr$

$=$ moment of R about O.

Cor. The algebraical sum of the moments of any number of coplanar forces which meet at a point about a point on the line of action of their resultant is zero.

4·21. The Moment of a Force about a Line is defined to be the product of the resolved part of the force at right angles to the line and the shortest distance between the force and the line.

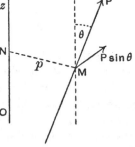

Thus if P be a force and Oz be a line which does not intersect P, $MN = p$ the shortest distance between P and Oz, and θ the angle between P and a line through M parallel to Oz, then $P \sin \theta$ is the resolved part of P at right angles to Oz and $Pp \sin \theta$ is the moment of P about Oz.

If P intersects the line Oz or is parallel to Oz, then the moment of P about Oz is zero, because in the one case $p = 0$ and in the other $\sin \theta = 0$.

4·22. *If through any point A on the line of action of a force P a plane xOy be drawn at right angles to any line Oz meeting it in O, then the moment of P about Oz is equal to the moment about O of the projection of the force P on the plane xOy.*

Let $NM = p$ be the shortest distance between P and Oz, and let DM

parallel to Oz make an angle θ with P and meet the plane xOy in D.
Then the projection of the force P
on the plane xOy is a force $P\sin\theta$
acting in the line AD, and OD is
the perpendicular from O to AD,
so that the moment about O of
the projection of P on the plane
xOy

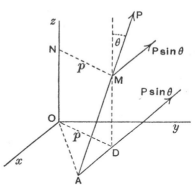

$$= OD \times P\sin\theta$$
$$= NM \times P\sin\theta = Pp\sin\theta$$
$$= \text{moment of } P \text{ about } Oz.$$

4·23. *When two forces act at a
point the algebraical sum of their
moments about any line is equal to
the moment of their resultant about
this line.*

Let P, Q be the forces acting at A, R their resultant and Oz the given
line.

Through A take a plane xOy at right angles to Oz and let P', Q', R'
denote the projections of the forces on the plane xOy.

Then the algebraical sum of the moments of P and Q about Oz is
equal to the algebraical sum of the moments of P' and Q' about O (**4·22**)

$$= \text{moment of } R' \text{ about } O \quad (\textbf{2·41 and 4·2}),$$
$$= \text{moment of } R \text{ about } Oz.$$

4·3. Parallel Forces. Let two forces P, Q acting at a point
O have a resultant R, and let any
straight line intersect the lines of action
of the forces in A, B and C.

We may use the centroid method as
in **2·72** and **3·43** to express the relations
between P, Q, R and the lines in the
figure, or we may proceed as follows:

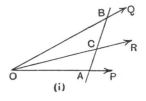

(i)

Take moments about A; then since P has no moment about
A, we have

$$\text{mt. of } Q \text{ about } A = \text{mt. of } R \text{ about } A$$

or $$Q.AB\sin B = R.AC\sin C$$

or $$Q.AB.OC = R.AC.OB$$

or $$\frac{Q}{OB.AC} = \frac{R}{OC.AB}.$$

Similarly by taking moments about B, we find that

$$\frac{P}{OA.CB} = \frac{R}{OC.AB},$$

therefore $$\frac{P}{OA.CB} = \frac{Q}{OB.AC} = \frac{R}{OC.AB} \quad \dots\dots\dots\dots(1).$$

Now let the point O move to an infinite distance while A, B remain fixed and the magnitude of the forces P, Q remain unaltered. Then OA, OB, OC become parallel lines, so that the resultant of two parallel forces is parallel to its components. Also since the ratios $OA : OB : OC$ tend to equality, therefore (1) becomes

$$\frac{P}{CB} = \frac{Q}{AC} = \frac{R}{AB} \quad \dots\dots\dots\dots\dots\dots(2).$$

But $$AB = AC + CB,$$

therefore $$R = P + Q \quad \dots\dots\dots\dots(3).$$

Figure (ii) shews what (i) becomes when O moves to an infinite distance. The direction in which O moves becomes the direction of the parallel forces but it does not affect the relations (2) and (3).

Next consider the case in which the transversal ABC cuts P on the other side of O as in fig. (iii). By the same argument as before we obtain the relation

$$\frac{P}{OA.BC} = \frac{Q}{OB.AC} = \frac{R}{OC.AB} \quad \dots\dots\dots\dots(4),$$

where BC, AC, AB are all positive lengths measured in the same sense.

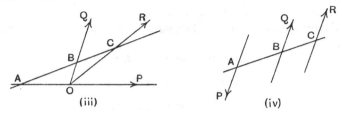

(iii)   (iv)

Then keeping A, B and the magnitudes of P, Q fixed as before, let O move to an infinite distance in any convenient direction so that P, Q, R become parallel as in fig. (iv) (where the direction in which O is moved in (iii) is along BO).

As before, since the ratios $OA : OB : OC$ tend to equality, (4) becomes

$$\frac{P}{BC} = \frac{Q}{AC} = \frac{R}{AB} \quad \dots\dots\dots\dots\dots(5).$$

But $AB = AC - BC$,

therefore $R = Q - P \quad \dots\dots\dots\dots\dots(6).$

Parallel forces are said to be *like* when they act in the same direction as in fig. (ii), and *unlike* when, as in fig. (iv), they act in opposite directions.

It follows from (3) and (6) that, if parallel forces are counted as positive or negative according to the sense in which they act, then the resultant of two parallel forces is equal to their algebraical sum; and from (2) and (5) it follows that the resultant divides AB in the ratio $Q : P$, internally or externally according as P and Q are like or unlike. Fig. (iv) shews the case $Q > P$; if however $P > Q$ the point C will lie on BA produced and R will act in the same sense as P.

4·31. Couples. The method of **4·3** fails to determine a resultant for two equal unlike parallel forces, for when $Q = P$ and the forces are unlike, (6) gives $R = 0$ and (5) gives $BC = AC$, implying that the point C is at an infinite distance. Such a pair of equal parallel forces acting in opposite senses is called a *couple*. A *torque* is another name for a couple. We shall discuss the properties of couples later in this chapter.

4·4. The theorem of moments of **4·2** can now be extended to include the case of parallel forces. For this purpose we shall prove that *the algebraical sum of the moments of two parallel forces about any point in their plane is equal to the moment of their resultant about the same point.*

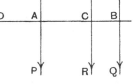

Let O be any point in the plane of the parallel forces P, Q. From O draw a line $OACB$ at right angles to P and Q, meeting

them in A, B and their resultant R in C. Then the algebraical sum of the moments of P and Q about O

$$= P.OA + Q.OB$$
$$= P(OC - AC) + Q(OC + CB)$$
$$= (P + Q)OC, \quad \text{since } P.AC = Q.CB,$$
$$= R.OC$$
$$= \text{moment of } R \text{ about } O.$$

This proof can be adapted for the case of unlike parallel forces, and for the case in which O lies between the forces.

It follows by the process of addition, step by step, that *the algebraical sum of the moments of any number of coplanar parallel forces about any point in their plane is equal to the moment of their resultant about the same point*; so that the theorem of **4·2** now holds good for all coplanar forces that have a resultant.

4·41. Since parallel lines project orthogonally into parallel lines and the ratio of lengths on a line is unaltered by orthogonal projection, it follows that the resultant of two parallel forces projects orthogonally into the resultant of their projections, and the same is true for the resultant of any number of parallel forces. Hence by the aid of **4·22** we deduce that *the algebraical sum of the moments about a line of any number of parallel forces is equal to the moment of their resultant about the same line.*

For, with the figure of **4·22**, the algebraical sum of the moments about Oz of parallel forces of which P is the type = algebraical sum of moments about O of the projections of the forces on the plane xOy = moment about O of the resultant of the projections of the forces = moment about O of the projection of the resultant force = moment about Oz of the resultant force.

4·5. Centre of Parallel Forces. Let n parallel forces $P_1, P_2, \dots P_n$ act at points $A_1, A_2, \dots A_n$. Then, whether the forces are all in the same sense or not, if we compound them together step by step, the resultant of P_1 and P_2 is $P_1 + P_2$ acting at a point B_1 on $A_1 A_2$ such that $P_1 . A_1 B_1 = P_2 . B_1 A_2$; next the resultant of $P_1 + P_2$ at B_1 and P_3 at A_3 is $P_1 + P_2 + P_3$ acting at a point B_2 on $B_1 A_3$ such that $(P_1 + P_2) B_1 B_2 = P_3 . B_2 A_3$ and so on. But this process is that defined in **2·6** as finding the centroid of the points $A_1, A_2, \dots A_n$ for multiples $P_1, P_2, \dots P_n$, and as proved in **2·61** it leads to a unique point G through which the resultant of the parallel forces P_1, P_2, ... P_n acts, save in the exceptional case in which

$$P_1 + P_2 + \ldots + P_n = 0 \qquad \ldots\ldots\ldots\ldots\ldots(1).$$

The vanishing of the algebraical sum of the parallel forces does not necessarily imply that the resultant is zero, because the forces may be equivalent to a couple. In order to determine whether this is so or not, we may divide the forces into two groups according as they represent positive or negative terms in (1). The resultants of the two groups are then equal in magnitude and opposite in sign, and because neither of these separate sums is zero we can use the centroid method above to determine definite points on the lines of action of these resultants of the two groups of forces. The system is thus either in equilibrium or equivalent to a couple according as the two resultants are found to act in the same straight line or not.

When the system is not equivalent to a couple, the unique point G determined as above is called the **centre of parallel forces**. Its position depends only on the positions of the chosen points A_1, A_2, ... A_n and the relative magnitudes of the forces. It does not depend on the direction of the parallel forces, so that the forces can be turned about their points of action remaining parallel to one another, and so long as they remain parallel the point G remains fixed.

4·51. Centre of Gravity. The weights of the particles which compose a material body constitute a set of forces directed towards the Earth's centre and therefore parallel with sufficient accuracy for our purpose. Since they are *like* parallel forces they have a resultant equal to their sum and this is called the *weight of the body* and it acts through a definite point in the body, viz. the centre of the parallel forces, which is called the *centre of gravity* or *centre of mass* of the body, or its *centroid*.

As thus defined the weight of a body is a force acting through a definite point of the body irrespective of the orientation of the body in reference to surrounding objects; and in writing down relations between the forces acting on a body the weights of its elements may in general be replaced by a single force acting at the centre of gravity

4·52. Centrobaric Bodies. When we speak of the weight of a body we mean 'the force with which the Earth attracts it'. In accordance with Newton's Law of Gravitation 'every particle in the universe attracts every other particle with a force directly proportional to the product of their masses and inversely as the square of the distance between them'. Newton proved that the resultant force of attraction exerted upon an external particle by the particles of a uniform sphere is the same as if the whole mass of the sphere were collected into a particle at its centre, and that the forces of attraction exerted upon the particles of a uniform sphere by an external body have a resultant passing through the centre of the sphere. But this does not hold good for bodies in general of various shapes. It follows that there is nothing approximative in the statement that the weight of a uniform sphere acts through its centre. It is an exact statement, and such a body is called *centrobaric* in that it has a real centre of gravity, but bodies in general only possess centres of gravity subject to the approximation that the Earth's attractions on the particles which compose the body may be regarded as parallel forces.

4·53. Analytical Formulae for Centre of Parallel Forces.

(i) *Coplanar forces.* Take rectangular axes Ox, Oy in the plane and let (x_1, y_1), (x_2, y_2), ... be the co-ordinates of the points of application A_1, A_2, ... of the parallel forces P_1, P_2, ... and $(\bar{x}, \bar{y})$ the co-ordinates of the centre G. Since the position of G does not depend on the direction of the parallel forces, we

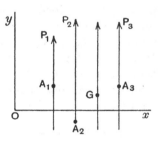

may suppose them to act parallel to the axis Oy, then by taking moments about O, we get

$$(P_1 + P_2 + P_3 + \dots)\bar{x} = P_1 x_1 + P_2 x_2 + P_3 x_3 + \dots$$

or
$$\bar{x} = \Sigma(Px)/\Sigma P.$$

Similarly, by taking the forces parallel to Ox, we get

$$\bar{y} = \Sigma(Py)/\Sigma P.$$

(ii) *When the forces are not coplanar.* Let (x_1, y_1, z_1), (x_2, y_2, z_2), ... be the co-ordinates of the points of application A_1, A_2, ... and $(\bar{x}, \bar{y}, \bar{z})$ the co-ordinates of the centre G of parallel forces P_1, P_2, P_3, ... referred to rectangular axes Ox, Oy, Oz.

Since the position of G is independent of the direction of the parallel forces, we may suppose them to be parallel to Oz, then by taking moments about Oy and using 4·41 we get

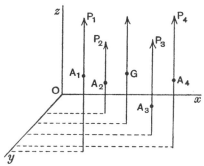

$$(P_1 + P_2 + P_3 + \ldots)\bar{x}$$
$$= P_1 x_1 + P_2 x_2 + P_3 x_3 + \ldots$$

or $\bar{x} = \Sigma(Px)/\Sigma P.$

Similarly

$$\bar{y} = \Sigma(Py)/\Sigma P,$$

and $\bar{z} = \Sigma(Pz)/\Sigma P.$

The method only fails to give a result when $\Sigma P = 0$, i.e. when the forces are either in equilibrium or equivalent to a couple.

4·54. *Position of equilibrium of a heavy body suspended from a given point without any other constraint.*

The body is in equilibrium under the action of two forces (i) its weight which acts vertically through its centre of gravity and (ii) the reaction at the point of support. Hence in the position of equilibrium the line joining the centre of gravity to the point of support must be vertical.

4·541. Examples. (i) *A uniform triangular lamina ABC is suspended from the corner A, and in equilibrium the side BC makes an angle α with the horizontal. Prove that*

$$2\tan\alpha = \cot B \sim \cot C. \qquad \text{[T.]}$$

It can be shewn that the centre of gravity of a uniform triangular lamina is at the intersection of the medians. Hence the vertical through A bisects BC in D.

Let AE be perpendicular to BC, then $EAD = \alpha$ and

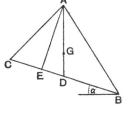

$$2\tan\alpha = 2ED/EA$$
$$= (EB - CE)/EA$$
$$= \cot B - \cot C.$$

(ii) *A heavy uniform rod of length $2a$ turns freely about a pivot at a point in it, and suspended by a string of length l fastened to the ends of the rod hangs a smooth bead of equal weight which slides on the string. Prove that the rod cannot rest in an inclined position unless the distance of the pivot from the middle point of the rod is less than a^2/l.* [T.]

Let AB be the rod and C the ring. Then since the bead is smooth the tensions in both parts of the string are the same so that CA and CB must be equally inclined to the vertical CD, or the vertical CD bisects the angle ACB. The weights of the rod and the ring are equal and parallel forces and they are balanced by the reaction of the pivot O, therefore we must have $GO = OD$, where G is the middle point of the rod.

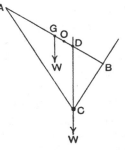

Then because CD bisects the angle ACB therefore

$$\frac{AD}{AC} = \frac{DB}{CB}, \quad \text{or} \quad \frac{a+GD}{AC} = \frac{a-GD}{CB},$$

therefore

$$\frac{2GD}{AC-CB} = \frac{2a}{AC+CB},$$

or

$$GD = \frac{a}{l}(AC - CB).$$

But $AC - CB$ must be less than AB or $2a$, therefore GO or $\frac{1}{2}GD$ must be less than a^2/l.

4·55. When a heavy body is placed upon a horizontal plane or on a plane inclined to the horizontal and sufficiently rough to prevent sliding, we may determine whether the body will stand or fall over as follows:

If the part of the body in contact with the plane has a convex boundary line—such as a triangle or a convex polygon—take this as 'the boundary of the base'; and in other cases, e.g. if the body were a chair, tie a string tightly round the parts of the body in contact with the plane, e.g. the feet of the legs of the chair, and take the line of the string as defining 'the boundary of the base'. The weight of the body acts vertically downwards through its centre of gravity and in equilibrium must be balanced by the pressures of the plane on the body, and however these pressures are distributed their resultant could not act·outside the boundary of the base. Hence it is necessary for equilibrium that the vertical through the centre of gravity shall intersect the plane within the boundary of the base.

4·6. Couples. Couples play an important part in the general theory of systems of forces and we shall now establish some of their principal properties.

Since a couple consists of two equal and opposite parallel forces, the algebraical sum of the resolved parts of the forces in every direction is zero, so that there is no tendency for the couple to produce in any direction a displacement of translation of the body upon which it acts; and, as we saw in **4·31**, the couple cannot be replaced by a single force.

The effect of a couple must therefore be measured in some other way, and, since it has no tendency to produce translation, we next consider what tendency it has to produce rotation.

Let the couple consist of two forces of magnitude P. It is of course assumed that they are both acting upon the same rigid body. Let us take the algebraical sum of the moments of the forces about any point O in their plane as the measure of their tendency to turn the body upon which they act about the point O.

From O draw a line at right angles to the forces meeting them in A and B.

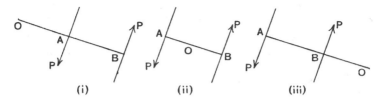

(i)   (ii)   (iii)

Then the algebraical sum of the moments about O (taken in the counter-clockwise sense in the three figures, which illustrate all cases) is

in fig. (i) $P.OB - P.OA = P.AB,$

in fig. (ii) $P.OB + P.AO = P.AB,$

and in fig. (iii) $P.AO - P.BO = P.AB.$

Thus the measure of the tendency of the couple to cause rotation about a point O is independent of the position of O in the plane and is measured by the product of either force and the distance between the forces; this distance is called the arm of the couple and the product is called the **moment of the couple**, and it may be counted as positive or negative according to the sense in which the couple tends to cause rotation.

4·61. The conclusion arrived at in the preceding article is of great importance. We may summarize it thus: the only way of measuring the effect of a couple is by considering its tendency to cause the rotation of a body; this tendency can be expressed by taking moments about a point, and the result obtained is independent of the point chosen and is a definite constant for each couple, namely the product of either force and the distance between the forces, which product we call the moment of the couple.

Further, the effect of a couple does not depend upon the absolute measure of its forces, but only upon its moment, i.e. the product force × distance; and if forces are measured in pounds and distances in feet, a couple is measured in foot-pounds.

When a body is acted upon by a couple and we consider its tendency to turn about a point in the plane of the couple, we have seen that the effect of the couple is independent of the relative positions of the point and the couple and depends only on the 'moment of the couple'; therefore *couples of equal moments in the same plane are equivalent*, and *couples of equal and opposite moments in the same plane balance one another*.

4·62. The equivalence of couples of equal moment is important because it means that we can represent a couple of given moment in a given plane by forces of any magnitude we like to choose and place one of them to act at any chosen point of the plane in any assigned direction, provided that we place the equal parallel force at the distance necessary to make a couple of the prescribed moment.

Thus a force and a couple in the same plane can always be reduced to a single force. For let P be a force acting at O and G the moment of a counter-clockwise couple, then we can re-present G by two forces of magnitude P at a distance p apart such that $Pp = G$, and place them so that one of them balances the given force P acting at O, and then the resultant of the given force P and the 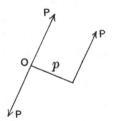 couple G is the single force P parallel to the given force and at a distance G/P from it.

4·621. Example. *Make use of* **4·62** *to find the resultant of two parallel forces.*

Let P, Q be the given parallel forces and let a line at right angles to them cut them in A and B.

At A introduce two equal and opposite forces of magnitude Q in the same straight line as P. This does not affect the resultant of the system,

but we may now consider the system to consist of a force $P + Q$ acting
at A and a couple of moment $Q . AB$ (clockwise in fig. (i)). This couple
can then be represented by equal and opposite forces of magnitude
$P + Q$, one of which can be placed at A to balance the former force $P + Q$

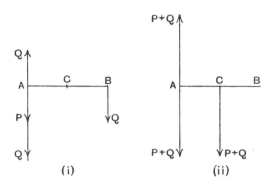

(i)   (ii)

acting at A (fig. (ii)), and the other will then cut AB at a point C such
that
$$(P + Q) AC = Q . AB,$$
or
$$P . AC = Q . CB,$$

and the resultant of the original forces P at A and Q at B is this force
$P + Q$ at C.

4·63. *Couples of equal moment in parallel planes are equi-*
valent.

Consider a couple of forces P at the ends of an arm AB. In
any parallel plane take a line $A'B'$ equal and parallel to AB,

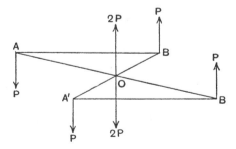

and let AB' and BA' intersect at O. The effect of the given
couple is not altered by introducing at O two equal and
opposite forces of magnitude $2P$ parallel to the forces of the

given couple; but we can now compound P at A and $2P$ (upwards) at O into a single force P at B', since

$$AB' : OB' = 2P : P.$$

Similarly P at B and $2P$ (downwards) at O have a resultant P at A'. We are therefore left with a couple of forces P at the ends of the arm $A'B'$ in a plane parallel to that of the original couple, so that any couple is equivalent to a couple of equal moment in any parallel plane.

4·64. Specification of a Couple. In order to specify a couple it is therefore only necessary to know (i) the direction of the set of parallel planes in any one of which the couple may be supposed to act, (ii) the magnitude of its moment, (iii) the sense in which it acts (clockwise or counter-clockwise).

These three properties of a couple can be represented completely by drawing a straight line

(i) at right angles to the set of parallel planes to indicate their direction,

(ii) of a measured length to indicate the magnitude of the moment,

(iii) in a definite sense to indicate the sense of the couple in accordance with a convention to be stated.

For example, a measured length MN on a certain straight line will represent a couple whose moment on a certain scale is of magnitude MN acting in a plane at right angles to MN in a certain sense, it being understood that a measured length NM on the same straight line will represent a couple of equal numerical moment but acting in the opposite sense. It is open to us to adopt any convention of signs; e.g. a clockwise 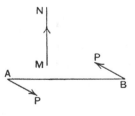 couple in the plane of the paper may be represented by a line at right angles to the paper on either side, i.e. towards or from the eye of the observer provided that a counter-clockwise couple is represented by a line drawn in the opposite sense.

A line drawn to represent a couple is called **the axis of the couple.**

4·641. A convenient method of specifying the sign of a couple is to take a set of three mutually perpendicular axes Ox, Oy, Oz and agree that a couple in the plane xOy which would cause rotation from Ox towards Oy shall be represented by a positive length along Oz. Similarly a couple in the plane yOz which would cause rotation from Oy towards Oz will be represented by a positive length along Ox, and a couple in the plane zOx which would cause rotation from Oz towards Ox will be represented by a positive length along Oy.

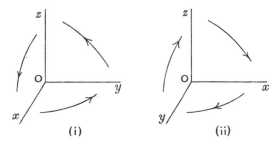

(i)   (ii)

Figures (i) and (ii) represent 'right-handed' and 'left-handed' sets of axes respectively, so-called because if what is known as a 'right-handed screw' were placed along Oz with its head at O in (i), it would have to be twisted in the sense of rotation from Ox towards Oy to drive it along Oz.

4·65. Composition of Couples. A couple is a vector. We have seen that a couple can be represented by its axis, i.e. a straight line of definite length in a definite direction; in order to shew that a couple is a vector it only remains to prove that couples can be compounded by applying the vector law of addition to their axes.

The composition of couples in the same or in parallel planes is effected by the algebraical addition of their moments, because we can choose a common arm for all the couples and we then have only to compound by algebraical addition the sets of forces which act at the ends of this arm. In terms of the axes of the couples the composition is effected by the addition of lengths measured on the same straight line.

When the couples are not in the same plane nor in parallel planes, consider two couples in planes which intersect. We may place the forces anywhere we please in the given planes and take them to be of any convenient magnitude provided that we adjust the arms of the couples suitably to make couples of given moments. Take a length AB on the line of inter-section of the planes together with a parallel length CD to represent the forces of one couple, and let the other couple be represented by forces BA and EF; and let AD, AE at right angles to AB be the arms of the couples.

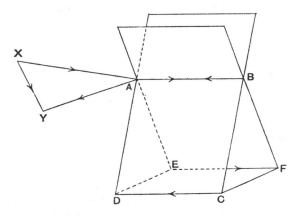

Then since the forces AB and BA balance one another, the resultant couple consists of the forces CD and EF, with an arm DE at right angles to both. Since the forces of all three couples are equal, the lengths of their axes will be proportional to their arms. Let XA equal to AD and at right angles to the plane $ABCD$ represent the first couple, and AY equal to AE and at right angles to the plane $BAEF$ represent the second couple. Then the triangle XAY is equal to the triangle DAE turned through a right angle in its own plane, so that XY is equal to DE and at right angles to the plane $CDEF$, and therefore XY represents the resultant couple on the same scale as XA and AY represent the component couples; and since the vector XY is the sum of the vectors XA and AY, therefore the axes of couples obey the vector law of addition.

It follows that when couples are represented by their axes they may be compounded and resolved in the same way as forces, and that any formula that has been established for the resultant of forces holds good for the resultant of couples. There is, however, an important distinction, that a force is *a vector localized in a line*, but a couple is a *non-localized vector* and all equal and parallel straight lines can be used indifferently to represent the same couple.

4·66. *Three forces represented in magnitude, direction and position by the sides of a triangle taken the same way round are equivalent to a couple.*

Let the lines AB, BC, CA represent the forces. By the vector law of addition the forces CA, AB have a resultant CB *acting at* A. Hence if AD be drawn equal and parallel to CB, then the couple BC, AD is equivalent to the three given forces. Since the arm of the couple

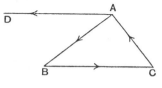

is the perpendicular from A to BC, its moment is represented by twice the area of the triangle.

4·67. *Forces completely represented by the sides of a plane polygon taken the same way round are equivalent to a couple whose moment is represented by twice the area of the polygon.*

This theorem follows from the last, if the polygon is divided up into triangles by joining one corner to each of the others, and equal and opposite forces are introduced represented by the lines so drawn. By this means we obtain a number of sets of three forces acting round triangles to each of which the last theorem applies, and by adding the equivalent couples we obtain

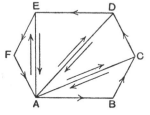

a single couple whose moment is, represented by twice the area of the polygon.

4·68. Examples. (i) *A stairway is made of n equal uniform rectangular blocks placed on top of each other and each projecting the same distance at the back beyond the one below. The top block is supported from below at its outermost point. Shew that the stairway can stand without mortar if, and only if, $2l > (n-1)a$, where $2l$ is the width of each block and a is the width of the tread.* [T.]

Let r blocks counting downwards from the top be supported by a force P acting at A, the outermost point of the top block, and a force

Q between the rth block and the $(r+1)$th, which must act either at B or to the left of B in the figure, where BC is the top of the $(r+1)$th block. If this condition holds whatever number of blocks we take, then the staircase will stand and not otherwise.

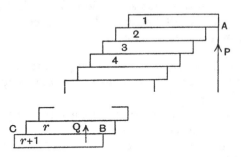

If x is the distance of Q from A, we must therefore have
$$x \geqslant ra \quad \dots\dots\dots\dots\dots\dots\dots\dots\dots(1).$$
But by moments about A
$$Qx = W\{l + (l+a) + (l+2a) + \dots + (l + (r-1)a)\}$$
$$= W\{rl + \tfrac{1}{2}ar(r-1)\} \quad \dots\dots\dots\dots\dots\dots\dots(2).$$
But $P + Q = rW$, and, by considering the top block alone, P is not greater than $\tfrac{1}{2}W$: for P is at a distance l from the middle point of the block, and, if P were greater than $\tfrac{1}{2}W$, the other force supporting the block would have to act at a distance greater than l from the middle point of the block which is impossible.

Therefore $\qquad\qquad P \leqslant \tfrac{1}{2}W,$

and $\qquad\qquad Q \geqslant (r - \tfrac{1}{2})W.$

Hence, from (2),
$$\{rl + \tfrac{1}{2}ar(r-1)\}/x \geqslant r - \tfrac{1}{2},$$
or $\qquad rl + \tfrac{1}{2}ar(r-1) \geqslant x(r - \tfrac{1}{2}) \quad \dots\dots\dots\dots\dots(3).$

But from (1) the necessary and sufficient condition for equilibrium is
$$x \geqslant ra;$$
hence from (3) $\qquad rl + \tfrac{1}{2}ar(r-1) \geqslant ar(r - \tfrac{1}{2}),$

or $\qquad\qquad l + \tfrac{1}{2}a(r-1) \geqslant a(r - \tfrac{1}{2}),$

or $\qquad\qquad 2l \geqslant ra.$

This condition is to hold good for all values of r from 1 to $n-1$, therefore the staircase stands if, and only if,
$$2l > (n-1)a.$$

(ii) *Two spheres, whose radii are a_1, a_2, rest inside a smooth hollow vertical right cylinder, of which the external radius is c, and the internal radius is b, where $b < a_1 + a_2$. Prove that, if the sphere a_1 is the lower, the* *cylinder will not overturn if its weight exceeds $w(2b-a_1-a_2)/c$, where w is the weight of the upper sphere.* [T.]

Since the cylinder is smooth the reactions between it and the spheres are horizontal forces, and the only vertical forces acting on the spheres are their weights and the reaction of the plane on which the lower one stands. This reaction passes through the point of contact and though it is equal to the sum of the weights w and w' of the spheres it cannot balance their resultant, so the vertical forces form a couple, viz. w downwards through the centre of the upper sphere and w upwards in the vertical through the centre of the lower sphere. The moment of this couple is $w(2b-a_1-a_2)$.

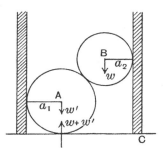

This is a couple acting on the spheres and if they remain in equilibrium this couple must be balanced by another, therefore the mutual reactions between the spheres and the cylinder constitute a couple of the same moment.

Considering next the equilibrium of the cylinder, when it is just about to overturn the reaction of the horizontal plane upon it will pass through a point C on its outer edge and form with the weight W of the cylinder a couple of moment Wc tending to keep the cylinder upright, and the cylinder will not overturn if the moment of this couple exceeds that of the other couple which acts in the contrary sense, i.e. if

$$W > w(2b-a_1-a_2)/c.$$

(iii) *$ABCD$ is a skew quadrilateral and forces are completely represented by the lines AB, BC, CD, DA; prove that they are equivalent to a couple of moment*

$AC \times BD \times$ sin (angle between AC and BD).

We have $\overline{AB}+\overline{BC}=\overline{AC}$ acting at B and $\overline{CD}+\overline{DA}=\overline{CA}$ acting at D. These two forces AC at B and CA at D form a couple whose arm is

$BD \times$ sin (angle between AC and BD),

and hence the required result.

(iv) *A body of weight W is suspended by two equal threads AP, BQ; the points of support A, B are on the same level at a distance a apart; and the threads are fastened to two points P, Q on the body so that PQ is horizontal and $PQ=b$. A couple G is applied about a vertical axis and the body is deflected through an angle θ with PQ at a depth h below AB. Shew that*

$$G = \tfrac{1}{4}W(ab/h)\sin\theta.$$ [T.]

Let MN be the projection of AB on the horizontal plane through PQ, and O the middle point of MN and PQ. Let T denote the tension in either thread and ϕ its inclination to the vertical. The weight W is supported by the vertical components of the tensions, therefore

$$W = 2T \cos \phi.$$

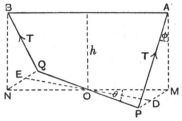

The couple G must balance the horizontal components of the tensions which act along PM and QN, therefore

$$G = T \sin \phi \times DE,$$

where DOE is perpendicular to PM and QN.

But $OD \cdot PM = 2\, \text{area}\, POM = \tfrac{1}{2}a \cdot \tfrac{1}{2}b \sin \theta,$

and $PM = h \tan \phi,$

therefore $DE = \tfrac{1}{2}(ab/h) \sin \theta \cot \phi,$

and $G = \tfrac{1}{2}T \cos \phi (ab/h) \sin \theta = \tfrac{1}{4}W (ab/h) \sin \theta.$

EXAMPLES

1. A uniform beam 20 ft. long and weighing 100 lb. is supported at its ends. If weights of 30 lb., 40 lb. and 50 lb. are placed on the beam at distances of 6 ft., 9 ft. and 15 ft. from one end, find the pressure on each support.

2. Two weights of 10 lb. and 2 lb. hang from the ends of a uniform lever a yard long and weighing 4 lb. Find the point in the lever about which it will balance.

3. $ABCD$ is a straight line, $AB = 2$ ft., $BC = 4$ ft., $CD = 3$ ft. Forces of 2 lb. and 6 lb. act vertically upwards at A and C, and forces of 7 lb. and 4 lb. act vertically downwards at B and D. Find the magnitude and line of action of the resultant.

4. A loaded horizontal plank 8 ft. long rests on supports at its ends. A load of 20 lb. is removed from a point 1 ft. from one end of the plank and placed at the middle. By how much is the pressure on each support altered.

5. A uniform rod AB, of weight W, rests horizontally on props at M, N, where $AM = \tfrac{1}{4}AB$, $AN = \tfrac{3}{5}AB$. The rod remains at rest when weights P and Q hang from A and B. Prove that

$$6P + W > 4Q > 2P - W.$$

Prove that, if $Q = P$, an additional vertical force which, applied at

B, will disturb equilibrium cannot be less than $\frac{1}{4}(2P+W)$, whether it act upwards or downwards.

6. A uniform rod AB, 6 ft. long, weighing 10 lb., can turn freely about a hinge at A. It carries a weight of 7 lb. at B, and is kept in equilibrium by a horizontal string CD, which is attached to a point D of the rod 5 ft. from A and to a point C 3 ft. vertically above A. Find the tension of the string.

7. Prove that if four forces acting along the sides of a square are in equilibrium they must be equal in magnitude.

8. A uniform ladder 13 ft. long, weighing 20 lb., rests against a smooth vertical wall with its lower end on rough ground 5 ft. from the bottom of the wall. Prove that the pressure of the ladder on the wall is a force of $4\frac{1}{6}$ lb.

9. Two horizontal wires inclined to one another at an angle 2α are attached to the top of a vertical post movable about its lower end, which is fixed. The post is supported by a stay inclined at an angle β to the vertical and fastened to a point two-thirds of the way up the post. Find the tension in the stay when the tensions in the wires are both T. Find also the horizontal and vertical components of the reaction at the lower end of the post.

10. A gate is supported by two hinges in a vertical line at a distance 3 ft. apart. The breadth of the gate is 5 ft. and its weight is 100 lb. The upper hinge exerts a horizontal force only and will yield when this force exceeds 250 lb. If a boy weighing 140 lb. stands on the gate without the hinge yielding, what is the greatest possible distance of his centre of gravity from the line of the hinges?

11. Three like parallel forces P, Q, R act at the corners of a triangle ABC; prove that their resultant passes through the circumcentre of the triangle for all directions of the forces if P, Q, R are in the ratios

$$a\cos A : b\cos B : c\cos C. \qquad [\text{C.}]$$

12. A, B, C are fixed points and D is on a fixed straight line. Find the locus of the centre of four equal parallel forces acting at A, B, C, D respectively. [S.]

13. A table, whose surface is horizontal, is in the shape of an equilateral triangle ABC, having a leg at each corner. If we neglect the weight of the table itself, find where a weight must be placed in order that the pressures on the legs may be as $1:2:3$. [S.]

14. A uniform rectangular board $ABCD$ of given weight is supported in a horizontal position on three pegs placed at the corner B and the middle points of the sides AD and CD. Find the pressures on the pegs, and prove that they are independent of the form of the rectangle. [S.]

4-2

15. A light horizontal beam, freely jointed at O, is supported and loaded as shewn. Determine the reactions at the supports.

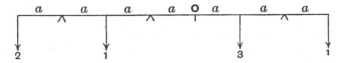

Determine the greatest weight which does not destroy equilibrium no matter where it is placed on the beam.   [S.]

16. Parallel forces of 1, 2, 3 lb. weight act at the corners A, C, E of a regular hexagon $ABCDEF$, and forces 4, 5, 6 lb. weight act at the corners B, D, F in a direction parallel to the first three but in the opposite sense. Find the point of application of their resultant.   [S.]

17. Forces of 1, 2, 3, 4, 5, 6 lb. weight respectively act at the corners of a regular hexagon inscribed in a circle of radius R, in directions perpendicular to the plane of the circle and in the same sense. Shew that the resultant cuts the plane at a distance $\dfrac{2R}{7}$ from the centre.   [S.]

18. A uniform rectangular brick 9 in. by $4\frac{1}{2}$ in. by 3 in. is placed on an inclined plane which is so rough that sliding does not take place, the longest edges being horizontal and the shortest perpendicular to the plane. Find the greatest angle to which the plane can be tilted without the brick toppling.

19. A uniform solid right circular cone stands on an inclined plane of angle $30°$. Prove that the cone will upset if its height exceeds $4\sqrt{3}$ times the radius of its base.

[The distance of the centre of gravity of such a cone from its base is one-quarter of its height.]

20. Two uniform rods AB, BC, rigidly jointed at B so that ABC is a right angle, hang freely in equilibrium from a fixed point at A. The lengths of the rods are a and b and their weights are wa and wb. Prove that, if AB makes an angle θ with the vertical,

$$\tan\theta = \frac{b^2}{a^2 + 2ab}.$$

21. The ends A, B of a light rod AB are joined by light inextensible strings AO, BO to a fixed point O, AO, BO being equal in length and at right angles to one another. Weights W_1 and W_2 are suspended from A and B. Shew that the rod will take up an inclination θ to the horizontal, where

$$\tan\theta = \frac{W_1 \sim W_2}{W_1 + W_2}.$$ [T.]

22. A sphere of weight W and radius a is suspended by a string of length l from a point P, and a weight w is also suspended from P by a

string sufficiently long for the weight to hang below the sphere. Shew
that the inclination of the first string to the vertical is

$$\sin^{-1}\frac{wa}{(W+w)(a+l)}.$$ [T.]

23. A point P is taken inside a triangle ABC, AP cuts BC in D.
Equal parallel forces in the plane of the triangle act at A, B, C, P, and
an equal but opposite parallel force acts at D. Find the position of P
so that the resultant may always pass through P, independently of the
direction of the forces. [T.]

24. P, Q are two like parallel forces. If two equal and opposite
forces S in any two parallel lines at a distance b apart in the plane of
P, Q are combined with them, shew that the resultant is displaced a
distance $bS/(P+Q)$. [T.]

25. Forces act along the sides BC, CA, AB of a triangle. Shew that
they reduce to a couple only if the forces are proportional to the sides.

26. If a couple of given moment and three forces in its plane are in
equilibrium, shew how the magnitudes of the forces can be obtained,
graphically or otherwise, when their lines of action are known. [I.]

27. If forces completely represented by the sides of a triangle taken
in order are in equilibrium with three equal forces acting at the corners
of the triangle along the tangents to the circumcircle the same way
round, prove that the triangle must be equilateral. [S.]

28. Three forces act along the sides of a triangle ABC taken in order
and are represented by these sides on a certain scale. They are in
equilibrium with three other forces acting along the sides of the pedal
triangle. Prove that these forces must be represented by the sides of
the pedal triangle on a scale which is to the first as

$$2\cos A\cos B\cos C:1.$$ [T.]

29. Forces λa, λb, λc parallel to the sides of a triangle ABC act at
A', B', C', the centres of the escribed circles: shew that they are
equivalent to a couple of moment $2\lambda(a+b+c)R$, where R is the radius
of the circumscribed circle. [S.]

30. $ABCD$ and $A'B'C'D'$ are two coplanar parallelograms. If
forces act along AA', $B'B$, CC', $D'D$ proportional to these respective
lengths, shew that they reduce to a couple, and find its moment. [S.]

31. Three forces λa, λb, λc act along AO, BO, CO respectively, where
O is the orthocentre of the triangle ABC. If they are rotated through
the same angle θ about A, B, C respectively, shew that they become
equivalent to a couple whose moment is $4\lambda\Delta\sin\theta$, where Δ is the area of
the triangle ABC. [S.]

32. Three coplanar forces are represented by straight lines AA',
BB', CC'. Shew that the forces will reduce to a couple if the c.g.'s of
the triangles ABC, $A'B'C'$ are the same, and if the straight lines are
not concurrent at a finite or infinite distance. [S.]

33. AB, $A'B'$ are two equal lines in the same plane, C and C' their middle points. Prove that forces represented by AA' and BB' are equivalent to a force represented by $2CC'$ and a couple whose moment is $2AC^2$ multiplied by the sine of the angle between AB and $A'B'$.

[S.]

34. P, Q, R are taken on the sides BC, CA, AB of a triangle, dividing each in the same ratio $1+\lambda : 1-\lambda$ in the same sense round the triangle. Prove that forces represented by AP, BQ, CR are equivalent to a couple whose moment is $2\lambda\Delta$, where Δ is the area of the triangle.   [S.]

35. Three uniform heavy rods OA, OB, OC mutually at right angles are rigidly connected at O. Their lengths are a, b, c respectively and they are suspended freely from O. Find the angle that OA makes with the vertical.   [S.]

36. A circular tray of radius a stands on a single circular foot of radius b. If w is the whole weight of the tray and its support, find how far from the centre a weight W can be placed without the tray falling over.   [S.]

37. Two smooth spheres each of radius r and weight W are placed inside a hollow cylinder of radius a, open at both ends, which rests on a horizontal plane with its axis vertical: prove that, in order that the cylinder may not upset, its weight must at least be

$$2W\left(1-\frac{r}{a}\right), \quad r > \tfrac{1}{2}a. \qquad [\text{S.}]$$

38. A uniform bar of weight W and length $2a$ is suspended, from two points in a horizontal plane, by two equal strings of length l, which are originally vertical: shew that the couple, which must be applied to the bar in a horizontal plane, to keep it at rest at right angles to its former direction is $\dfrac{Wa^2}{\sqrt{l^2-2a^2}}$.   [S.]

39. A body of weight W is suspended by two equal threads of length l, fastened to two fixed points, in the same horizontal line distant $2a$ apart and also tied to two fixed points in the body, distant $2b$ apart. Calculate the couple required to turn the body through an angle θ; and prove that when l is large compared with a or b, the couple is $W\left(ab/l\right)\sin\theta$.   [C.]

40. Two straight uniform beams ACD, CDB of weights W_1 and W_2 are braced together at C and D and the compound beam thus formed is supported at its ends A and B. Prove that the bracings have to sustain tensions and thrusts equal respectively to

$$\frac{1}{2}\frac{AD(BD.W_1+BC.W_2)}{BC.AD-BD.AC} \text{ and } \frac{1}{2}\frac{BC(AD.W_1+AC.W_2)}{BC.AD-BD.AC}.$$

Prove that, if the weights of the beams are inversely proportional to their lengths, these forces constitute a torque acting on each beam. [T.]

41. Equal lengths AA', BB' are marked off in the same direction along a given straight line, and equal lengths CC', DD' along another given line. Prove that forces completely represented by AC, $C'A'$, CB, $B'C'$, BD, $D'B'$, DA, $A'D'$ are in equilibrium.

42. $OABC$ is a tetrahedron. Shew that couples represented by the areas OBC, OCA, OAB can have a couple represented by the area ABC as resultant.

43. $ABCD$ is a skew quadrilateral and forces represented completely by AB, BC, CD, DA act in those lines; give a geometrical construction to determine the moment of the system about any line and prove that the moments about any two lines are as the cosines of the inclinations of those lines to the common perpendicular to AC and BD.

[C.]

ANSWERS

1. 114·5 lb., 105·5 lb. **2.** 9 in. from one end. **3.** 3 lb., 4 ft. 8 in. from A. **4.** $7\frac{1}{2}$ lb. **6.** 19·2 lb. **9.** $3T \cos \alpha \operatorname{cosec} \beta$; $T \cos \alpha$, $3T \cos \alpha \cot \beta$. **10.** $3\frac{4}{7}$ ft. **12.** A parallel at three-fourths of the distance from the centroid of A, B, C. **13.** Take D on BC so that $BD = \frac{3}{5}BC$, and G on AD so that $AG = \frac{5}{8}AD$. **14.** One-third of the weight on each peg. **15.** 6 lb. **16.** If M, N are the middle points of AF and CD, G divides MN in the ratio $7:11$. **18.** 56° 19′. **23.** The middle point of the median through A. **30.** The moment may be represented by the area

$$ABCD - A'B'C'D' - 2ABB' + 2DCD'.$$

35. $\cos^{-1}\{a^2/\sqrt{(a^4 + b^4 + c^4)}\}$. **36.** $b(1 + w/W)$. **39.** $(Wab \sin \theta)/\sqrt{(l^2 - a^2 - b^2 + 2ab \cos \theta)}$.

Chapter V

COPLANAR FORCES

5·1. *A system of coplanar forces acting on a rigid body can be reduced to a single force acting at an arbitrarily chosen point in the plane of the forces together with a couple.*

Let O be the chosen point and P any one of the forces at a distance p from O. At O introduce a pair of equal and opposite forces equal and parallel to P. This does not affect the body, but the three forces P may now be regarded as a force P at O, parallel to the original force P and in the same sense, together with a couple of moment Pp. Hence any force P can be transferred to act at O parallel to its 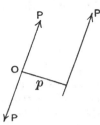 original direction by the introduction of a suitable couple whose moment is the moment of P about O.

Let all the forces of the given system be transferred to act at O in this way by the introduction of suitable couples. Then the forces acting at O can be compounded into a single force R and the couples can be compounded into a single couple of moment G by the algebraical addition of their moments. Also R is the resultant of all the given forces moved parallel to themselves to act at O, and G is the algebraical sum of their moments about O.

5·11. Since R is the resultant of all the forces moved parallel to themselves to act at a point, it follows that R has the same magnitude and direction no matter at what point O in the plane it acts. But G is the algebraical sum of the moments of the forces about O and therefore depends in general upon the position of O in the plane.

5·12. *When neither R nor G is zero* (fig. (i)) the system can be reduced to a single force R. For the couple can be represented by two forces R at a distance G/R apart and so placed

that one of them balances R at O (fig. (ii)) as in **4·62**. The
system has then been reduced
to a single force R acting at a
point O' at a distance G/R
from O.

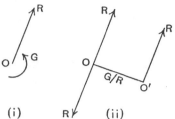

5·13. *When $G = 0$ and R is
not zero.* The system is then
equivalent to a single force R,
or the chosen point O is a point
about which the algebraical sum of the moments of the forces
is zero. It follows that when a system of coplanar forces has a
single resultant the algebraical sum of the moments of the
forces about any point on the line of action of the resultant is
zero.

5·14. *When $R = 0$ and G is not zero.* The system is then
equivalent to a couple G. Also the vanishing of R does not
depend upon the choice of O, and therefore in this case G is
independent of O or the algebraical sum of the moments of the
forces about every point in the plane is the same.

5·15. *When $R = 0$ and $G = 0$* the system is in equilibrium.

5·2. *Conditions of equilibrium of a system of coplanar forces.*
When the system of forces is reduced to a single force R at
an arbitrarily chosen point O and a couple G, it is necessary
and sufficient for equilibrium that $R = 0$ and $G = 0$.

These conditions can be expressed in other ways, thus:

(i) Since R represents the resultant of all the forces moved
parallel to themselves to act at a point, R will vanish if and
only if the algebraical sums of the resolved parts of the forces
in two perpendicular directions in the plane are separately
zero. Again $G = 0$ means that the algebraical sum of the
moments of the forces about the point O is zero. Hence it is
necessary and sufficient for equilibrium that *the algebraical sums
of the resolved parts of the forces in two perpendicular directions
in the plane and of the moments of the forces about a point in the
plane should separately be zero.*

(ii) From **5·13** it follows that when the algebraical sum of the moments of the forces about a point is zero if there is a resultant force it must pass through this point. Hence another way of stating the conditions of equilibrium is that *the algebraical sums of the moments of the forces about three non-collinear points O, O', O'' should separately be zero.* For the vanishing of the sum of the moments about O means that $G = 0$, and if there is a resultant R it must pass through O; but in like manner it must pass through O' and O'' and this is not possible because O, O' and O'' are not collinear, therefore $R = 0$.

(iii) If the algebraical sums of the moments of the forces about two points O and O' are zero, then $G = 0$, and if there is a resultant R it must pass through O and O'. This latter possibility will be excluded if we know that the algebraical sum of the resolved parts of the forces in any direction not perpendicular to OO' is zero, for a force along OO' would have a component in every direction save perpendicular to OO'. It follows that the system of forces is in equilibrium if *the algebraical sums of the moments of the forces about two points and of the resolved parts of the forces in any direction not at right angles to the line joining the points are separately zero.*

5·3. Analytical Method. The conclusions that we have established so far in this chapter can also be obtained analytically as follows:

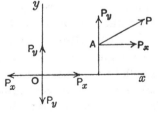

Let P be any one of a system of coplanar forces which acts at a point A whose co-ordinates are x, y referred to rectangular axes in the plane. Let P_x, P_y be the components of P parallel to the axes Ox, Oy. At O introduce equal and opposite forces P_y along the axis of y. One of these forces together with P_y at A forms a couple of moment xP_y in the counter-clockwise sense in the figure. Similarly by introducing at O equal and opposite forces P_x along the axis of x and combining one of these forces with P_x at A we get a couple yP_x in the clockwise sense. The two couples are equivalent to a single couple $xP_y - yP_x$ in the sense of rotation from Ox

towards Oy, and we have in addition the forces P_x, P_y at O parallel to their original directions. If we proceed in the same way with all the forces of the given system and compound together by addition the forces along Ox and along Oy and add the couples, we get a force at O with components

$$X = \Sigma P_x \text{ along } Ox$$

and
$$Y = \Sigma P_y \text{ along } Oy,$$

and a couple
$$G = \Sigma (xP_y - yP_x).$$

The forces X, Y can be compounded into a single force R given by
$$R^2 = X^2 + Y^2,$$

and this force is clearly the resultant of all the given forces moved parallel to themselves to act at O, while G is the algebraical sum of the moments about O of the forces.

It is evident that G depends upon the choice of origin O and that R does not.

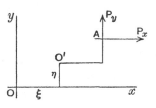

If we reduced the system to a force R at a point O' of coordinates (ξ, η) instead of at O, we can deduce the value of the corresponding couple G' from the formula obtained for G by subtracting ξ, η from x, y respectively, so that

$$G' = \Sigma \{(x - \xi) P_y - (y - \eta) P_x\}$$
$$= \Sigma (xP_y - yP_x) - \xi \Sigma P_y + \eta \Sigma P_x,$$

or
$$G' = G - \xi Y + \eta X.$$

Again, since the algebraical sum of the moments about any point on the line of action of the resultant is zero, O' is on the resultant if
$$G - \xi Y + \eta X = 0,$$

and, if we regard ξ, η as current co-ordinates, this equation is the equation of the line of action of the resultant force.

5·31. Conditions of Equilibrium.

(i) It is clearly necessary and sufficient for equilibrium that
$$R = 0 \text{ and } G = 0,$$

or that
$$X = 0, \quad Y = 0 \text{ and } G = 0.$$

These conditions expressed in words are the same as those of **5·2** (i).

(ii) An equivalent set of conditions is that the algebraical sums of moments of the forces about each of three non-collinear points should be zero. For if the sums of the moments about O, O', O'' are zero, we may write

$$G = 0,$$
$$G' \equiv G - \xi Y + \eta X = 0$$

and
$$G'' \equiv G - \xi' Y + \eta' X = 0,$$

where ξ, η are the co-ordinates of O', and ξ', η' those of O''.

Since O, O', O'' are not collinear, therefore $\eta/\xi \neq \eta'/\xi'$ and the equations can only be satisfied by $X = 0$ and $Y = 0$ so that these conditions, i.e. (ii) of **5·2**, are equivalent to (i).

(iii) Lastly, conditions (iii) of **5·2** are represented by equations

$$G = 0,$$
$$G' \equiv G - \xi Y + \eta X = 0$$

and
$$X = 0,$$

taking the direction of resolution to be the axis of x and the line OO' not perpendicular to Ox, i.e. ξ is not zero. The equations then require that $Y = 0$, and so the conditions are equivalent to (i).

5·32. Examples. (i) *ABCDEF is a regular hexagon and O is its centre. Forces of magnitudes* 1, 2, 3, 4, 5, 6 *act in the lines AB, CB, CD, ED, EF, AF in the senses indicated by the order of the letters. Reduce the system to a force at O and a couple, and find the point in AB through which the single resultant passes.*

Take the axis Ox along BE, and a perpendicular axis Oy as in the figure, then using the notation of **5·3** we have

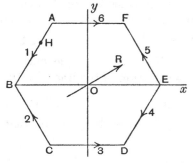

$$X = -1.\tfrac{1}{2} - 2.\tfrac{1}{2} + 3 - 4.\tfrac{1}{2}$$
$$-5.\tfrac{1}{2} + 6$$
$$= 3,$$

and

$$Y = -1.\frac{\sqrt{3}}{2} + 2.\frac{\sqrt{3}}{2} - 4.\frac{\sqrt{3}}{2}$$
$$+ 5.\frac{\sqrt{3}}{2}$$
$$= \sqrt{3}.$$

Hence the single force at O is R, where $R^2 = 9 + 3 = 12$, so that $R = 2\sqrt{3}$, and it makes with Ox an angle $\tan^{-1}\dfrac{Y}{X}$, i.e. $\tan^{-1}\dfrac{1}{\sqrt{3}}$ or $30°$.

Again $G = $ sum of moments about O of the forces
$$= (1 - 2 + 3 - 4 + 5 - 6)\,a$$
$$= -3a,$$

in the counter-clockwise sense, where a is the distance of O from a side of the hexagon. The couple is therefore 3 in the clockwise sense and it can be represented by two forces of magnitude R so placed that one of them balances R at O and the other, the single resultant, is parallel to R and at a distance G/R, i.e. $3a/2\sqrt{3}$ or $\sqrt{3}a/2$, from O. It is easily seen that the line of action passes through the middle points of AB and AF.

Here we should note that there is a much more direct method of finding the point in AB through which the resultant passes, if that fact and nothing more about the resultant were required. Thus let H be the required point, and let $AH = z$. Then, since H is a point through which the resultant passes, the algebraical sum of the moments of the forces about H is zero; therefore

$6z \sin 60° - 5\,(a + z \sin 60°) + 4 \cdot 2a - 3\,(2a - z \sin 60°)$
$$+\, 2\,(a - z \sin 60°) = 0,$$

or $\sqrt{3}z - a = 0.$

Therefore $AH = a/\sqrt{3} = \tfrac{1}{2}AB.$

(ii) *AB, CD are segments of two fixed coplanar lines. If AB be of fixed length and likewise CD, shew that the resultant of forces represented by AC, DB has a fixed direction and magnitude.*   [S.]

We have a force

$\overline{AC} = \overline{AB} + \overline{BC}$ acting at A,

and $\overline{DB} = \overline{DC}' + \overline{CB}$ acting at D.

Then the forces BC at A and CB at D
constitute a couple and no matter what

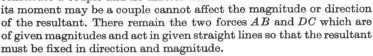

its moment may be a couple cannot affect the magnitude or direction of the resultant. There remain the two forces AB and DC which are of given magnitudes and act in given straight lines so that the resultant must be fixed in direction and magnitude.

(iii) *Forces P, Q, R act along the sides BC, CA, AB of a triangle ABC, and forces P', Q', R' act along OA, OB, OC, where O is the centre of the circumscribing circle. Prove that, if the six forces are in equilibrium,*

$$P \cos A + Q \cos B + R \cos C = 0,$$

and $$\dfrac{PP'}{a} + \dfrac{QQ'}{b} + \dfrac{RR'}{c} = 0.$$ [S.]

Since the angle BOC is $2A$, therefore the distance of the force P from

O is $OB \cos A$, with similar expressions for the distances of the forces Q and R. Hence, by taking moments about O for the six forces, we get

$$P \cos A + Q \cos B + R \cos C = 0.$$

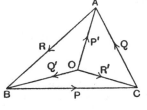

Again the moment about A of a force represented by BC is twice the area ABC, therefore the moment about A of the force P along BC is $\dfrac{P}{BC} \times 2 \triangle ABC$. Hence, by taking moments about A, we get

$$\frac{P}{BC} \times 2 \triangle ABC + \frac{R'}{OC} \times 2 \triangle AOC - \frac{Q'}{OB} \times 2 \triangle AOB = 0.$$

Similarly, by taking moments about B and C, we get

$$\frac{Q}{CA} \times 2 \triangle ABC + \frac{P'}{OA} \times 2 \triangle BOA - \frac{R'}{OC} \times 2 \triangle BOC = 0,$$

and

$$\frac{R}{AB} \times 2 \triangle ABC + \frac{Q'}{OB} \times 2 \triangle COB - \frac{P'}{OA} \times 2 \triangle COA = 0.$$

Now multiply these equations respectively by P', Q', R' and add, and we get the result

$$\frac{PP'}{a} + \frac{QQ'}{b} + \frac{RR'}{c} = 0.$$

(iv) *If a system of forces in one plane reduces to a couple whose moment is G and when each force is turned round its point of application through a right angle reduces to a couple H, prove that when each force is turned through an angle α the system is equivalent to a couple whose moment is*

$$G \cos \alpha + H \sin \alpha. \qquad \text{[S.]}$$

Let any force P of the system act at the point (x, y) and make an angle θ with the axis Ox. The components of P are $P \cos \theta$ and $P \sin \theta$ and the sum of their moments about the origin O is

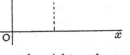

$$xP \sin \theta - yP \cos \theta.$$

Hence, since the system reduces to a couple G, we must have

$$G = \Sigma (xP \sin \theta - yP \cos \theta).$$

In like manner, when each force is turned through a right angle, we have

$$H = \Sigma \{xP \sin (\theta + 90°) - yP \cos (\theta + 90°)\}$$
$$= \Sigma (xP \cos \theta + yP \sin \theta);$$

and when each force is turned through an angle α, the moment of the couple is

$$\Sigma \{xP \sin (\theta + \alpha) - yP \cos (\theta + \alpha)\}$$
$$= \cos \alpha \, \Sigma (xP \sin \theta - yP \cos \theta) + \sin \alpha \, \Sigma (xP \cos \theta + yP \sin \theta)$$
$$= G \cos \alpha + H \sin \alpha.$$

(v) *Prove that, if a variable system of forces in a plane have constant moments about two fixed points in the plane, the resultant passes through another fixed point and its minimum value is the difference of the moments divided by the distance between the two points.* [S.]

Let O, O' be the two fixed points about which the moments are given constants G, G'. Take OO' as axis of x and a perpendicular through O as the axis of y, and let O' be the point $(\xi, 0)$. Then as in **5·3**, if X, Y are components of the resultant force,

$$G' = G - \xi Y \quad \ldots\ldots\ldots(1).$$

But G, G' and ξ are all definite constants, therefore Y is a constant. Hence X is the only variable part of the system of forces. But the equation of the line of action of the resultant is

$$G - xY + yX = 0, \quad (5\text{·}3)$$

and this passes through the fixed point $\left(\dfrac{G}{Y}, 0\right)$ whatever be the value of X.

The minimum value of the resultant is got by taking $X = 0$, the resultant then being Y, and from (1) this is equal to $(G - G')/\xi$, or $(G - G')/OO'$, which proves the proposition.

EXAMPLES

1. $ABCD$ is a square of side a whose diagonals intersect at O. Forces 1, 2, 3, 4, $6\sqrt{2}$ act along AB, BC, CD, DA, AC. Find the magnitude and direction of the single force at O and the magnitude of the couple which together are equivalent to the given forces. Also find at what distance from A the resultant of the five forces cuts the line AB.

2. Forces 1, 3, 5, 7, $8\sqrt{2}$ act along the sides AB, BC, CD, DA and the diagonal BD of a square of side a. Taking AB and AD as axes of x and y respectively, find the magnitudes of the resultant force and the equation of its line of action.

3. $ABCD$ is a square of side 8 in. Forces 3, 2, 5, 4 act along AB, BC, DC, DA in the senses indicated by the order of the letters. Find the resultant force and the point in which it cuts CD.

4. ABC is an equilateral triangle of side a and forces 4, 2, 2 act along AB, AC, BC in the directions indicated by the letters. Find the resultant force and the distances from A at which it cuts AB and AC.

5. A force lies in the plane of the triangle ABC and its moments about A, B, C are L, M, N respectively. If the force is the resultant of three forces acting along the sides of the triangle, determine the magnitudes and senses of these forces.

6. Forces P, $2P$, $3P$, $4P$ act along the sides AB, BC, CD, DA respectively of the square $ABCD$. Find the magnitude of their resultant and the points in which it cuts the sides AB and BC.   [S.]

7. Three forces act along the sides of an equilateral triangle, and vary so that their resultant is constant in magnitude and direction. Shew that the difference of the magnitudes of any two of the three forces is constant.   [S.]

8. Forces P, $2P$, $3P$, $4P$, $5P$ act respectively along the lines AB, CA, FC, DF, ED, where $ABCDEF$ is a regular hexagon. Find the forces along the lines BC, FA, FE which will keep the system in equilibrium.   [S.]

9. Forces P_1, P_2, P_3, P_4, P_5, P_6 act along the sides of a regular hexagon taken in order. Shew that they will be in equilibrium if

$$\Sigma P = 0 \quad \text{and} \quad P_1 - P_4 = P_3 - P_6 = P_5 - P_2.$$ [S.]

10. ABC is an equilateral triangle and D, E, F are the middle points of the sides BC, CA, AB. Forces P, $2P$, $3P$ act along BC, CA, AB and forces $4P$, $5P$, $6P$ along FE, ED, DF. Find the line of action of the resultant.   [I.]

11. Three forces P, Q, R act along the sides of the triangle formed by the lines

$$x + y = 1, \quad y - x = 1, \quad y = 2.$$

Find the equation of the line of action of their resultant.   [I.]

12. Forces ka, kb, kc act along the sides a, b, c of a triangle, and a fourth force P is given in magnitude. Prove that, if the line of action of P touches a given circle of radius r, the line of action of the resultant will touch a fixed circle; and find its radius.   [S.]

13. Prove that a force acting in the plane of a triangle ABC can be replaced uniquely by three forces along the sides of the triangle.

If each of a system of coplanar forces be replaced in this way by forces of type $p.BC$, $q.CA$ and $r.AB$, shew that the necessary and sufficient conditions that the system reduces to a couple are that

$$\Sigma p = \Sigma q = \Sigma r.$$ [S.]

14. Forces P, Q, R act along the sides BC, CA, AB of a triangle. Find the condition that their resultant should be parallel to BC and determine its magnitude.   [S.]

15. Three forces each equal to P act along the sides of a triangle ABC in order; prove that the resultant is

$$P\left(1 - 8\sin\frac{A}{2}\sin\frac{B}{2}\sin\frac{C}{2}\right)^{\frac{1}{2}},$$

and find the distance of its line of action from one angular point, and where it cuts one side of the triangle.   [S.]

16. Forces P, Q, R act along the lines $x = 0$, $y = 0$, and
$$x \cos \theta + y \sin \theta = p.$$
Find the magnitude of the resultant and the equation of its line of action. (The axes of co-ordinates are rectangular.) [S.]

17. Forces P, Q, R act respectively along the lines BC, CA, AB; prove that, if S is the magnitude of their resultant,
$$S^2 = P^2 + Q^2 + R^2 - 2QR \cos A - 2RP \cos B - 2PQ \cos C,$$
and that the line of action of this resultant will pass through the ortho-centre of the triangle ABC, if
$$P \sec A + Q \sec B + R \sec C = 0. \qquad [S.]$$

18. Three forces P, Q, R act along the sides BC, CA, AB of a triangle ABC. Their resultant lies in the line joining the centre of the circle inscribed in ABC and the centre of gravity of the triangular lamina ABC. Shew that
$$P : Q : R = (b - c) a : (c - a) b : (a - b) c. \qquad [S.]$$

19. Forces P, Q, R acting along the sides BC, CA, AB of a triangle are in equilibrium with forces P', Q', R' acting along AG, BG, CG, the lines joining the vertices to the centre of gravity of the triangle. Prove that
$$\frac{PP'}{AG.BC} + \frac{QQ'}{BG.CA} + \frac{RR'}{CG.AB} = 0. \qquad [S.]$$

20. A system of forces X, Y, Z acting along the sides of the triangle ABC is equivalent to a system P, Q, R along the sides of the pedal triangle. Prove that $2P = Y/\cos B + Z/\cos C.$ [S.]

21. Three forces act at the angular points of a triangle ABC, each parallel to the opposite side. If they have a resultant acting along the line joining the in-centre and orthocentre of the triangle, prove that they are proportional to
$$ab (a + b) \cos B - ac (a + c) \cos C$$
and the two similar expressions. [S.]

22. Forces P, Q, R, S acting in the sides AB, BC, CD, DA of a quadrilateral $ABCD$ are in equilibrium: shew that
$$\frac{P \times R}{AB.CD} = \frac{Q \times S}{BC.DA}.$$
Will this condition alone ensure equilibrium? [S.]

23. Four forces in equilibrium act along the sides of a quadrilateral inscribed in a circle. Prove that each force is proportional to the sine of the angle in the segment of the circle cut off by the side opposite to its line of action. [S.]

24. Four forces act along the sides of a quadrilateral $ABCD$ and are represented in direction and magnitude by BA, BC, AD, CD. Prove that their resultant is parallel to one diagonal and bisects the other. [I.]

R S

5

25. $ABCD$ is a quadrilateral in which AD, BC are parallel. Prove that forces represented by AB, BC, CD, DA, AC, DB have a resultant of magnitude $2EF$, where E, F are the middle points of AD, BC, and find its position. [T.]

26. A lamina is acted on by four forces represented in magnitude, direction and position by the sides AB, BC, CD, DA of a crossed quadrilateral, the lines AB, CD crossing one another. If the lamina is pivoted at O, the point of intersection of AC and DB, and can turn in its plane, prove that it can be kept in equilibrium by a force in BC of magnitude $\dfrac{AC.BD}{BO.CO} BC$. [I.]

27. Four forces act in the sides AB, BC, CD, DA of a plane quadrilateral and are in equilibrium; shew that

$$X_{AB}/AB \triangle DAB \triangle ABC = X_{CB}/BC \triangle ABC \triangle BCD$$
$$= X_{CD}/CD \triangle BCD \triangle CDA = X_{AD}/AD \triangle CDA \triangle DAB. \quad [\text{C.}]$$

28. Prove that in all cases a system of coplanar forces can be replaced by two forces, one of which acts through a given point and the other along a given straight line. [S.]

29. A system of coplanar forces is such that the sum of their moments about any point in a given straight line is always the same. State what inferences may be drawn about the resultant of the system. [S.]

ANSWERS

1. $4\sqrt{2}$ along AC, $5a$; $5a/4$. 2. $4\sqrt{10}$; $x + 3y = 4a$.
3. $\sqrt{68}$; 20 in. from D. 4. $2\sqrt{7}$, $\frac{1}{2}a$, a.
5. $La/2\Delta$, $Mb/2\Delta$, $Nc/2\Delta$. 6. $2\sqrt{2}P$; $\frac{7}{5}AB$ from B.
8. $2\left(1 + \dfrac{1}{\sqrt{3}}\right)P$; $-(9 - 2\sqrt{3})P$; $-\left(11 - \dfrac{10}{\sqrt{3}}\right)P$.
10. Parallel to CB dividing DA in the ratio $1:5$.
11. $P(x + y - 1) + Q(y - x - 1) - R\sqrt{2}(y - 2) = 0$.
12. $r \pm 2k\Delta/P$. 14. $Qc = Rb$; $P - Qa/b$.
15. Distance from $A = 2P\Delta/aR$, where resultant R divides BC externally in the ratio $c:b$.
16. $\sqrt{(P^2 + Q^2 + R^2 - 2QR \sin\theta + 2RP \cos\theta)}$;
  $Px - Qy + R(x \cos\theta + y \sin\theta - p) = 0$.
25. R cuts DA produced in L so that $AL = FB$ and is parallel to EF.
29. R is either zero or parallel to the given line.

Chapter VI

THE SOLUTION OF PROBLEMS

6·1. We have found that conditions sufficient to ensure the equilibrium of a rigid body under the action of coplanar forces are three in number, and may be summarized as—

two resolutions and one equation of moments,

or   one resolution and two equations of moments,

or   three equations of moments.

The student will do well to bear in mind that, in the solution of a problem, no additional information can be obtained by writing down *more* than three such equations of moments or resolutions for one and the same body.

It follows that many statical problems are indeterminate or insolvable without further hypotheses as to the elastic properties of the body acted upon. For example, if a beam or bar under the action of given forces has its ends fixed, it is not possible to determine the magnitude and direction of the reactions at the fixed ends, because this means four unknown quantities, two magnitudes and two directions, and the conditions of equilibrium give only three equations between the four unknown quantities.

6·2. Constraints and Degrees of Freedom. The position of a plane body in its plane is determined by three quantities, such as the rectangular co-ordinates of one specified point of the body and the angle which a line fixed in the body makes with a line fixed in the plane. Hence a body free to move in a plane has *three degrees of freedom*. We may also express this fact by saying that it is free to have a motion of translation with components in either or both of two perpendicular directions and also free to rotate.

The number of degrees of freedom may be reduced by the imposition of constraints: thus for a body with two points fixed no motion is possible, i.e. it has *no degrees of freedom*; but with one point fixed, rotation is possible, i.e. the body has *one degree of freedom*; and if a point of the body is constrained to lie on a given line, then the point may move along the line and the body turn about the point so that it has *two degrees of freedom*.

In general the loss of a degree of freedom means the imposition of a constraining force. There are three equations of equilibrium and these constitute relations which should enable us to determine the three co-ordinates which specify the position of an unconstrained body in equilibrium under the action of given forces. The three equations should also enable us to determine the smaller number of co-ordinates of a body deprived of one or two degrees of freedom together with some of the elements of the constraint.

Thus the equations determine in the case of

complete freedom	three co-ordinates of position,
two degrees of freedom	two co-ordinates of position and one element of constraint,
one degree of freedom	one co-ordinate of position and two elements of constraint,
no freedom	three elements of constraint.

But as stated in **6·1** the fixing of two points in a body under the action of given forces imply *four* unknown elements, viz. the magnitudes and directions of two unknown constraining forces, so such problems are indeterminate unless some fact about one of the constraining forces is prescribed.

6·21. There is a large class of problems in which a body is in equilibrium under the action of three forces. We can shew that in such circumstances *the forces must be coplanar* and either concurrent or parallel.

We assume that since the forces are in equilibrium they can have no tendency to turn the body on which they act about any straight line, or that the algebraical sum of their moments about any straight line is zero. A formal proof of this statement will be found in **14·2**.

Let P, Q, R be the forces. Take any points A, B on the lines of action of P and Q. Then the three forces being in equilibrium the sum of their moments about AB is zero; but P and Q intersect AB and have no moment about it, therefore the moment of R about AB is zero. It follows that R is either parallel to AB or intersects it in a point C. We may reject the former alternative since the points A and B are chosen at random on the lines of action of P and Q. Hence AB intersects R in C. Similarly if D be any other point on the line of action of Q, AD must intersect

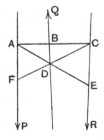

R in a point E. And for a like reason CD must intersect the force P in a point F. The three forces P, Q, R have therefore been proved to lie in the plane of the triangle ADC.

Again if no two of the forces are parallel, two of them, say P and Q, must meet in a point O, and since R is equal and opposite to the resultant of P and Q, therefore R also passes through O and the forces are concurrent. But if two of the forces are parallel, then the third force which balances the resultant of the first two is also parallel to them.

6·22. When the number of forces acting on a body in equilibrium is three, or can be reduced to three, the position of equilibrium can be determined by making use of the fact that these forces must be concurrent or parallel, the expression of this fact geometrically involving relations between the parts of the figure which serve to determine the position of equilibrium. This is called the geometrical method of solution. In the alternative analytical method we write down the three equations between the forces obtained from the conditions of equilibrium, i.e. by resolving and taking moments, and use these equations to obtain the required solution.

We will exemplify these methods by the solution of some problems.

6·23. Examples. (i) *A triangular lamina ABC is suspended from a point O by light strings fastened to the points A, B and hangs so that the side BC is vertical. Prove that, if α, β are the angles which the strings AO, BO make with the vertical, then*

$$2 \cot \alpha - \cot \beta = 3 \cot B.$$

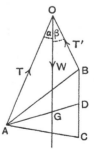

The three forces acting on the lamina are the tensions of the strings OA, OB and the weight. The weight must therefore act through O; but it is a vertical force, therefore OG is vertical, where G is the centre of gravity (intersection of medians) of the triangle.

Now, if AD is the median through A,

$$AG = 2GD,$$

and OG and BC are parallel, so that by projecting on the horizontal

$$AO \sin \alpha : BO \sin \beta = AG : GD = 2 : 1 \quad \ldots\ldots\ldots\ldots(1).$$

Also by projecting on the horizontal and vertical

$$AO \sin \alpha + BO \sin \beta = AB \sin B \quad \ldots\ldots\ldots\ldots(2),$$

and $\qquad AO \cos \alpha - BO \cos \beta = AB \cos B \quad \ldots\ldots\ldots\ldots(3).$

Then, by substituting from (1) in (2) and (3), we get

$$3BO \sin \beta = AB \sin B,$$

and $\qquad \dfrac{BO}{\sin \alpha}(2 \cos \alpha \sin \beta - \sin \alpha \cos \beta) = AB \cos B;$

therefore, by division $\quad 2 \cot \alpha - \cot \beta = 3 \cot B \quad \ldots\ldots\ldots\ldots\ldots(4).$

Alternatively, if we approach the problem analytically, we call the tensions T, T' and the weight W.

By resolving horizontally and vertically, we get

$$T \sin \alpha = T' \sin \beta \quad \ldots\ldots\ldots\ldots\ldots\ldots(5),$$

and $\qquad T \cos \alpha + T' \cos \beta = W \quad \ldots\ldots\ldots\ldots\ldots(6).$

We can also obtain information by taking moments about a point in the plane, and since T and T' pass through O we choose O for this purpose as leading to the simplest result. The result is that the moment of W about O is zero so that W must act through O. The reader should note that we cannot, by taking moments about any other point or resolving in any other direction, obtain any additional information than is contained in (5), (6) and the fact that W acts through O. Also that (5) and (6) suffice only to express T and T' in terms of W, α and β, and are equivalent to an application of Lami's Theorem

$$T : T' : W = \sin \beta : \sin \alpha : \sin (\alpha + \beta);$$

and that for the geometrical fact expressed by (4) we are compelled to resort to the geometrical solution given above.

(ii) *A uniform rod has a ring at one end which slides along a smooth vertical wire; the rod rests touching a smooth cylinder of radius r whose axis is horizontal and at a distance c from the wire. If θ is the angle the rod makes with the horizontal, find its length in terms of c, r, and θ.* [S.]

Let AB be the rod and G its middle point, XY the vertical wire and A the ring. It is assumed that the vertical plane through the rod cuts the cylinder in a circle of centre C and radius r. The rod touches this circle at H. The rod is in equilibrium under the action of three forces, namely, its weight acting vertically downwards through G, the re-action of the smooth cylinder along the radius through H, and the reaction of the smooth wire at right angles to the wire, i.e. horizontally through A, and these three forces must meet in a point O.

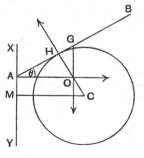

Then, if CM is perpendicular to XY and $2l$ denotes the length of the rod, we have $\quad c = CM = AO + CO \sin \theta = l \cos \theta + CO \sin \theta,$

and $\qquad l \sin \theta = OG = OH \sec \theta = (r - CO) \sec \theta;$

therefore, by eliminating CO

$$l\sin\theta\cos\theta = r - CO$$
$$= r - (c - l\cos\theta)\operatorname{cosec}\theta,$$

or $\qquad l\cos^3\theta = c - r\sin\theta.$

The reactions at A and H can then be expressed in terms of the weight by using Lami's Theorem.

We leave it to the reader to shew that there is another solution

$$l\cos^3\theta = c + r\sin\theta,$$

in which the horizontal reaction of the vertical rod on the ring is reversed in direction.

(iii) *$ABCD$ is a uniform lamina, in shape a rhombus with sides of length a and the angle $A = 2\alpha$. P and Q are smooth pegs, PQ being of length l and horizontal. Find the angle which AC makes with the vertical if the lamina can rest with points on the sides AB, AD in contact with the pegs and with AC not vertical. Shew that such a position of equilibrium occurs only if* $\qquad a\cos^2\alpha\sin\alpha < l < a\cos^2\alpha.$ [S.]

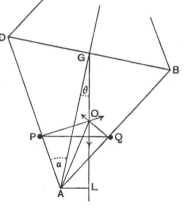

The reactions of the smooth pegs at P and Q are at right angles to AP and BQ and meet in O, and since the only other force acting on the lamina is its weight, the vertical GL through the centre G of the rhombus must pass through O. Let AG make an angle θ with the vertical in equilibrium; then DB makes an angle θ with the horizontal, and the angle

$$APQ = ADB + \theta,$$
and $\qquad AQP = ABD - \theta,$
therefore
$$APQ - AQP = 2\theta.$$

Since APO and AQO are right angles, therefore AO is the diameter of a circle $APOQ$, and

$$AG\sin\theta = AO\sin AOL = AO\sin(AOQ - QOL)$$
$$= AO\sin(APQ - AQP)$$
$$= AO\sin 2\theta.$$

Hence, when θ is not zero,

$$AG = 2AO\cos\theta.$$

But $AG = a\cos\alpha$, and, since AO is the diameter of a circle in which a chord $PQ (=l)$ subtends an angle 2α at the circumference, therefore

$$AO = l\operatorname{cosec} 2\alpha.$$

Hence $\qquad a\cos\alpha = 2l\operatorname{cosec} 2\alpha\cos\theta,$

or $\qquad l\cos\theta = a\cos^2\alpha\sin\alpha.$

It follows at once that there is such an angle θ, if and only if

$$a\cos^2\alpha\sin\alpha < l.$$

Again, if AL is a horizontal line through A, then GAL is the complement of θ, and GAQ is α, and AQ must be above AL so that

$$\alpha < \tfrac{1}{2}\pi - \theta,$$

or $$\sin\alpha < \cos\theta;$$

therefore $$l\sin\alpha < l\cos\theta,$$

i.e. $$l\sin\alpha < a\cos^2\alpha\sin\alpha,$$

so that $$l < a\cos^2\alpha,$$

i.e. another necessary condition.

6·3. When a problem relates to the equilibrium of two or more bodies in contact under the action of given external forces, we assume the existence of mutual actions and reactions at the points of contact of the bodies, and we also assume that when the bodies are smooth these forces act at right angles to the common tangent planes at the points of contact. We may then write down the equations of equilibrium for each body separately, including in such equations the given external forces which act.upon the body and the unknown reactions of the other bodies upon it. By eliminating the unknown reactions we then get a smaller number of equations connecting the given external forces and the co-ordinates of position and so arrive at a solution as in a problem of a single body. In some cases it is useful to write down the equations of equilibrium for two bodies regarded as though they were a single body and thus avoid the introduction of the unknown reaction between them. Sometimes both methods are combined in the solution of the same problem. It is essential, however, to remember that the force between two bodies is always a mutual action and reaction, and to keep clearly in mind in writing down an equation to what body or bodies it refers so that all the forces which act on the body or bodies are included and none others.

The examples of the next article will serve to illustrate the methods described above.

6·31. Examples. (i) *Two smooth spheres of equal weight rest inside a fixed smooth right circular cone which has its vertex downwards and its axis inclined at an angle β to the vertical, each sphere touching the cone at one* *point only. Prove that the common normal to the spheres makes an angle*

$$\cot^{-1}\left\{\frac{\sin 2\beta}{\cos 2\alpha + \cos 2\beta}\right\}$$

with the vertical, where α is the semi-vertical angle of the cone.   [S.]

A, B are the centres of the spheres and C, D their points of contact with the sides of the cone. They need not have the same radius, but they have the same weight, so that, if we regard the two as one body, their weight acts in the vertical through G the middle point of AB. There is a mutual action and reaction between the spheres along the line of centres AB, but, if we regard the two as forming one body, the only forces in addition to the weight are the pressures of the sides of the cone along the radii of the spheres at C and D.

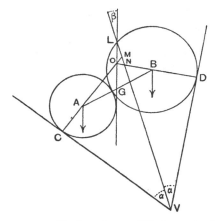

Since we are now considering the equilibrium of a body under the action of three forces these forces must be coplanar and concurrent, so that CA and DB meet in a point O, and O must lie on the vertical through G.

We are now in a position to obtain a geometrical solution as follows. Let GO, CO and DO meet the axis of the cone in L, M and N. Then the angle

$$OLM = \beta \quad \text{and} \quad OMN = ONM = \tfrac{1}{2}\pi - \alpha;$$

therefore

$$AOG = LOM = \tfrac{1}{2}\pi - \alpha - \beta,$$

and

$$BOG = OLN + ONL = \beta + \tfrac{1}{2}\pi - \alpha.$$

Let $OGB = \theta$, then

$$\frac{\sin AOG}{\sin OAG} = \frac{AG}{OG} = \frac{BG}{OG} = \frac{\sin GOB}{\sin OBG},$$

or

$$\frac{\cos(\alpha + \beta)}{\sin(\theta - \tfrac{1}{2}\pi + \alpha + \beta)} = \frac{\cos(\alpha - \beta)}{\sin(\theta + \beta + \tfrac{1}{2}\pi - \alpha)},$$

or

$$\frac{-\cos(\alpha + \beta + \theta)}{\cos(\alpha + \beta)} = \frac{\cos(\alpha - \beta - \theta)}{\cos(\alpha - \beta)}.$$

Therefore

$$-\cos\theta+\sin\theta\tan(\alpha+\beta)=\cos\theta+\sin\theta\tan(\alpha-\beta),$$

or
$$2\cot\theta=\tan(\alpha+\beta)-\tan(\alpha-\beta)$$
$$=\frac{\sin 2\beta}{\cos(\alpha+\beta)\cos(\alpha-\beta)};$$

whence
$$\theta=\cot^{-1}\left\{\frac{\sin 2\beta}{\cos 2\alpha+\cos 2\beta}\right\}.$$

(ii) *Inside a fixed hollow cylinder of radius R, whose generators are horizontal, there are placed symmetrically two equal cylinders, each of radius r: a third cylinder equal to each of the latter is placed symmetrically on them; shew that equilibrium cannot exist unless $R < r(1+2\sqrt{7})$.* [S.]

Taking a vertical section through the centres of the axes of the cylinders, let O, A, B, C be the centres and let W denote the weight of each of the smaller cylinders. The figure is symmetrical about the vertical OC. Let X denote the mutual action and reaction between the two lower cylinders, T the mutual reactions in the lines CA and CB, and S the pressures of the fixed cylinder on each of the two lower ones acting along AO and BO.

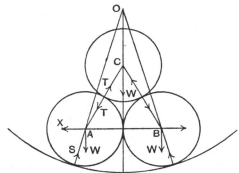

Let the angle $AOB = 2\theta$. The angle ACB is $60°$.

By considering the three smaller cylinders as one body and resolving vertically, the weight $3W$ is balanced by the vertical components of the forces S, therefore

$$2S\cos\theta=3W \quad\dots\dots\dots\dots\dots\dots\dots(1).$$

Then, resolving vertically for the upper cylinder alone, we get

$$2T\cos 30°=W, \quad\text{or}\quad \sqrt{3}T=W \quad\dots\dots\dots\dots\dots(2);$$

and, resolving horizontally for the cylinder centre A, we get

$$X+T\cos 60°=S\sin\theta \quad\dots\dots\dots\dots\dots(3).$$

Now equilibrium could not exist if X were negative, because X represents the pressure between the two lower cylinders; hence we must have

$$T\cos 60°\leqslant S\sin\theta.$$

On substituting for S and T in terms of W from (1) and (2), we get

$$\frac{W}{2\sqrt{3}} \leqslant \frac{3W}{2}\tan\theta,$$

or

$$\tan\theta \geqslant \frac{1}{3\sqrt{3}}.$$

But

$$\tan\theta = \frac{r}{\sqrt{\{(R-r)^2 - r^2\}}},$$

so the required condition becomes

$$\frac{r^2}{(R-r)^2 - r^2} \geqslant \frac{1}{27},$$

or

$$28r^2 \geqslant (R-r)^2,$$

or

$$R \leqslant r(1 + 2\sqrt{7}).$$

6·4. Reactions at Joints. There are a large number of problems in which two bodies are described as 'smoothly hinged' at a point. In such a case the hinge may be regarded as a pin passing through cylindrical holes in the bodies, closely fitting and so smooth that each body can turn about the pin without friction. When the hinge or joint is smooth the reaction of the pin on either body reduces to a single force, because, no matter how many points of contact there may be between the pin and the cylindrical hole in the body, the reaction at each of these points acts along the common normal and therefore passes through the centre of the pin (considering only forces in one plane) and all such forces can be combined into a single force through the centre of the pin. When the pin connects two bodies A and B only, then the pin is subject to two forces only, namely the reactions of A and B upon it, and in equilibrium these must be equal and opposite. But the reactions of the pin on the bodies are equal and opposite to the former forces, so that the result of the smooth joint is to set up equal and opposite forces on the bodies A and B and it is unnecessary to consider the precise form of the joint, because it is sufficient to know that, as the result of the smooth joint, there is a pair of equal and opposite forces between the bodies at a certain point and that the bodies are so constrained that the only possible relative motion is one of turning about this point.

6·41. Example. *Two uniform rods AB, BC of equal weight are smoothly hinged at B. The end A can turn about a fixed point and BC rests* *across a smooth horizontal peg. If in equilibrium both rods make angles of 60° with the vertical, prove that the reaction at B divides the angle ABC into angles whose tangents are as −1 : 14.* [S.]

The rod AB is acted upon by three forces, its weight W acting through its middle point G, the mutual reaction R at the hinge B, and a constraining force at A; and these three forces being in equilibrium must meet in a point D.

The rod BC is also acted upon by three forces, its weight W acting through its middle point H, the mutual reaction R at the hinge B, and the pressure of the smooth peg P acting at right angles to BC. These three forces must also meet in a point E.

The reader is advised to draw figures with the rods or reactions in other positions in order to convince himself that the position shewn in the figure is the only one in which the directions of the two sets of three forces acting at a point can possibly result in equilibrium, and in particular that it is not possible for B to be lower than A and C.

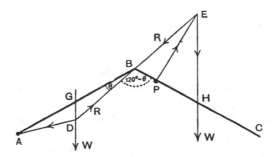

Let θ denote the angle ABD.

Then at D we have $\dfrac{W}{R} = \dfrac{\sin ADB}{\sin ADG}$, by Lami's Theorem, and at E we have $\dfrac{W}{R} = \dfrac{\sin BEP}{\sin PEH}$ for the same reason; therefore

$$\frac{\sin BEP}{\sin PEH} = \frac{\sin ADB}{\sin ADG} = \frac{AD \sin ADB}{AD \sin ADG}$$

$$= \frac{AB \sin ABD}{AG \sin AGD},$$

or $$\frac{\cos(60° + \theta)}{\sin 30°} = \frac{2 \sin \theta}{\sin 60°}.$$

Therefore $$\sqrt{3}\cos(60° + \theta) = 2 \sin \theta,$$

so that $$\tan \theta = \sqrt{3/7}.$$

Then $$\frac{\tan ABD}{\tan DBC} = \frac{\tan \theta}{\tan(120° - \theta)} = \frac{\tan \theta\,(1 - \sqrt{3}\tan \theta)}{-\sqrt{3} - \tan \theta}$$

$$= -\frac{1}{14}.$$

6·42. Working Rules. When more than two bodies are connected at the same joint, then it is necessary to exercise care in regard to the reactions because what is described as 'the reaction at the joint' usually depends on the way in which the attachments are made.

Before considering further problems we will state some working rules which serve to simplify the process in many cases.

(i) *When a framework and the forces acting upon it are both symmetrical about a line passing through a joint, then the reaction at that joint is perpendicular to the line of symmetry.*

This follows because no reason could be assigned for supposing the reaction to act on one side of the perpendicular which would not by symmetry also be a reason for supposing it to act on the other side of the perpendicular.

(ii) *When a body is in equilibrium under the action of forces acting upon it at two points A, B, only, then the forces at these points must act in the line AB and be equal and opposite.*

For otherwise the forces could not maintain equilibrium. This applies to all cases of rods which form parts of jointed frameworks and are acted upon by no forces save the reactions at the joints; the reactions in any such case must be along the rod.

In many problems a 'light' rod or a string is used to connect two points in order to limit the freedom in a framework. A 'light' rod means one whose weight is negligible.

Since the only possible effect of such a rod or string is to produce forces along its length at the points which it connects, we may frequently effect a simplification in the consideration of the nature of a joint by substituting for the rod or string two equal and opposite forces acting on the framework at the ends of the rod or string in its direction. We may treat such forces as external forces applied to the framework without affecting the solution of the problem.

We shall make some use of this artifice in the problems that follow.

6·43. Examples. (i) *Four rods are smoothly jointed at their extremities to form a quadrilateral ABCD, and the opposite corners A, C* *and B, D are joined by tight strings. Prove that, if T, T′ denote the tensions in AC and BD, then*

$$T\left(\frac{1}{AO}+\frac{1}{OC}\right)=T'\left(\frac{1}{BO}+\frac{1}{OD}\right),$$

where O is the intersection of the diagonals.

In obtaining the required result we can avoid all difficulties about the joints by regarding the strings as fastened to the rods AD and BC, and regarding the rods AB and CD as replaced by the reactions R_1, R_3 which they produce. The rod AD is then in equilibrium under the action of four forces T and R_1, at A, and T' and R_3 at D.

By taking moments about D, we get

$$T \cdot AD \sin CAD = R_1 \cdot AD \sin BAD.$$

Similarly by taking moments about C for the rod BC, we get

$$T' \cdot BC \sin CBD = R_1 \cdot BC \sin ABC.$$

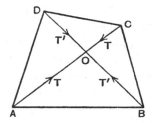
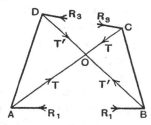

Therefore $T \sin ABC \sin CAD = T' \sin BAD \sin CBD.$

Using $\triangle ABC$ to denote the area ABC, this gives

$$\frac{T \cdot \triangle ABC \cdot \triangle CAD}{AB \cdot BC \cdot AC \cdot AD}=\frac{T' \cdot \triangle BAD \cdot \triangle CBD}{AD \cdot AB \cdot BC \cdot BD},$$

or

$$\frac{T}{AC} \cdot AC \cdot BO \sin O \cdot AC \cdot OD \sin O = \frac{T'}{BD} \cdot BD \cdot AO \sin O \cdot BD \cdot OC \sin O,$$

or

$$\frac{T \cdot AC}{AO \cdot OC}=\frac{T' \cdot BD}{BO \cdot OD},$$

i.e.

$$T\left(\frac{1}{AO}+\frac{1}{OC}\right)=T'\left(\frac{1}{BO}+\frac{1}{OD}\right).$$

It will be noticed that in this solution we have avoided altogether the question of the nature of the joints and the reactions at the joints. A simple method of representation would be to suppose the rods connected by pin joints and the strings attached to the pins. Each rod can then be supposed to end in a small ring which passes over the pin, and the forces act only along the rods and along the strings.

The figure shews these three forces at the corner B, and their ratios are given by Lami's Theorem.

This affords an alternative method of obtaining the relations between T, T' and R_1, used in the foregoing solution.

(ii) *ABCD is a rhombus formed by four light rods smoothly jointed at their ends and PQ is a light rod smoothly jointed at one end to a point P in BC and at the other end to a point Q in AD. Two forces each equal to F are applied at A and C in opposite directions along AC. Prove that the stress in PQ is* $F.AB.PQ/AC(AQ \sim BP)$.   [S.]

If the joints are regarded as pins fixed to the rods AD and BC passing through holes in the rods AB and CD, while the forces F are applied to the pins all the conditions are fulfilled and the part played by the rods AB and CD is that of exerting equal and opposite forces at their ends in the direction of their lengths.

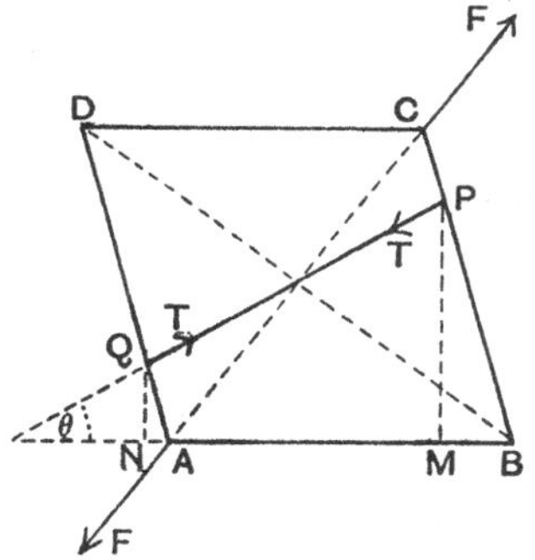

Let T denote the stress in PQ. Consider the equilibrium of the rod AD and resolve at right angles to AB; then

$$T \sin \theta = F \sin CAB,$$

where θ is the inclination of PQ to AB.

Then, if PM, QN are perpendicular to AB, we have

$$\sin \theta = \frac{MP \sim NQ}{PQ} = \frac{BP \sim AQ}{PQ} \sin ABC.$$

Therefore $T = \dfrac{F.PQ}{BP \sim AQ} \cdot \dfrac{\sin CAB}{\sin ABC} = \dfrac{F.PQ.AB}{AC(BP \sim AQ)}.$

6·44. In order to simplify the solutions of problems it is important to make use of symmetry wherever it exists and to introduce no more unknown reactions than is necessary.

The following problem could be solved by first inserting in the figure unknown horizontal and vertical reactions at every one of the six joints, but it will be noticed that in the solution given no unknown quantities are introduced save the two that are to be compared.

6·45. Examples. (i) *A hexagon ABCDEF is formed of six equal rods of the same weight W smoothly jointed at their extremities. It is suspended from the point A and the regular form is maintained by light rods BF and CE. Prove that the thrust in the former is five times that in the latter.*

Suppose that the rod BF is attached to the two upper rods and the rod CE to the two lower rods. Let R and S denote the thrusts in BF and CE. Then since the only effect of these rods is to produce thrusts at their ends, we may ignore these rods if instead of them we suppose horizontal forces R to act outwards on AB at B and on AF at F, and horizontal forces S to act outwards on CD at C and on DE at E. Begin by inserting these forces in the figure.

Then consider the equilibrium of the rod CD. The reaction at D is horizontal because there is symmetry about the vertical through D. But the only horizontal forces on CD are the force S at C and the reaction at D, so that this reaction at D must be equal and opposite to S. Then as regards vertical forces: the weight W acts vertically downwards through the middle point of CD and the only other vertical force can be at C, therefore there is a reaction at C which acts vertically upwards and is equal to W. Insert this in the figure; and, since it is produced by the rod CB, also insert an equal and opposite force W downwards acting at C on CB.

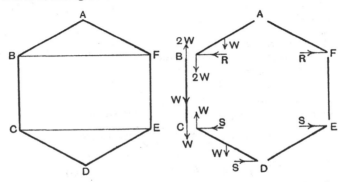

We can now express S in terms of W by taking moments for the rod CD about C or about D, or, what is the same thing, equating the moments of the two couples that act upon the rod. We find that

$$S.CD\sin 30° = W.\tfrac{1}{2}CD\cos 30°,$$

or $$S = W\sqrt{3}/2.$$

Then, returning to the figure, consider the rod BC. It is in equilibrium under the action of its weight W, a downward force W at C and the reaction at B. This latter force must therefore act vertically upwards and be equal to $2W$. Insert this force in the figure and also the equal and opposite reaction $2W$ at B on AB.

Then consider the rod AB. It is in equilibrium under the action of its weight W, the horizontal and vertical forces R and $2W$ at B and the reaction at A. It is not necessary to specify the latter because we can take moments about A; by so doing we find that

$$R.AB\sin 30° = \tfrac{5}{2}W.AB\cos 30°,$$

or $$R = 5W\sqrt{3}/2 = 5S.$$

If the reactions at A were required, their precise form would depend upon the method of support, but so far as our figure takes us, it is clear that AB must be acted upon at A by a horizontal force R towards AF, and by a vertical force $3W$; also that the total supporting force at A is equal to the total weight $6W$.

(ii) *A regular pentagon $ABCDE$ formed of five uniform rods, each of weight W, freely hinged to each other at their ends is placed in a vertical plane with CD resting on a horizontal plane and the regular pentagonal form is maintained by means of a string joining the middle points of the rods BC and DE. Prove that the tension in the string is*

$$\left(\cot\frac{\pi}{5} + 3\cot\frac{2\pi}{5}\right)W. \qquad \text{[S.]}$$

It is only necessary to consider the reactions at the corners A and B. By symmetry that at A is horizontal and equal say to X. The rod AB is also acted on by its weight W and the reaction at B. The latter must therefore have a horizontal component X and a vertical component W upwards. Insert in the diagram forces at B acting upon BC in the opposite senses. Then by taking moments about B for the rod AB, since the rod AB makes an angle $\frac{1}{5}\pi$ with the horizontal, we get

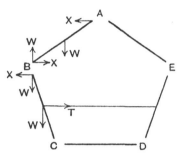

$$X\sin\frac{\pi}{5} = \tfrac{1}{2}W\cos\frac{\pi}{5} \qquad\dots\dots\dots\dots\dots\dots(1),$$

after dividing by the length of the rod.

Again if T denotes the tension in the string which joins the middle points of BC and DE, by taking moments about C for the rod BC, which makes an angle $\frac{2}{5}\pi$ with the horizontal, we get

$$\tfrac{1}{2}T\sin\frac{2\pi}{5} = \tfrac{1}{2}W\cos\frac{2\pi}{5} + W\cos\frac{2\pi}{5} + X\sin\frac{2\pi}{5} \quad\dots\dots\dots(2),$$

after dividing by the length of the rod.

On substituting for X in terms of W from (1), we find that

$$T = W\left(\cot\frac{\pi}{5} + 3\cot\frac{2\pi}{5}\right).$$

6·5. Chain of Heavy Particles. Let a number of heavy particles of weights $w_1, w_2, w_3, \dots$ be fastened to a light string the ends of which are held fixed. Let the successive portions of the string be inclined to the vertical at angles $\theta_1, \theta_2, \theta_3, \dots$ and let the tensions in these portions be denoted by $T_1, T_2, T_3, \dots$ as indicated in fig. (i).

By considering the equilibrium of the successive particles and resolving horizontally, we get

$$T_1\sin\theta_1 = T_2\sin\theta_2 = T_3\sin\theta_3 = \dots,$$

or the horizontal component of tension is constant and

$$= T \text{ say} \quad \dots\dots\dots\dots\dots\dots\dots\dots(1).$$

Then by resolving vertically for each particle we get

$$T_2 \cos \theta_2 - T_1 \cos \theta_1 = w_1,$$
$$T_3 \cos \theta_3 - T_2 \cos \theta_2 = w_2,$$
$$T_4 \cos \theta_4 - T_3 \cos \theta_3 = w_3,$$
$$\text{etc.,}$$

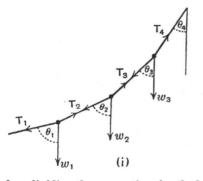

(i)

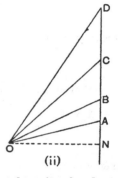

(ii)

and on dividing these equations by the horizontal tension they become

$$\cot \theta_2 - \cot \theta_1 = \frac{w_1}{T},$$

$$\cot \theta_3 - \cot \theta_2 = \frac{w_2}{T},$$

$$\cot \theta_4 - \cot \theta_3 = \frac{w_3}{T},$$

$$\text{etc.,}$$

so that if the weights are equal the cotangents of the inclinations to the vertical of the successive portions of string increase in arithmetical progression.

6·51. We can also treat the problem graphically by constructing triangles of forces for the three forces acting on each particle. Thus in fig. (ii) OBA is a triangle of forces for the particle w_1, in which BA represents w_1, AO represents T_1 and OB represents T_2; then OCB is a triangle of forces for the particle w_2, in which CB represents w_2, BO represents T_2 and OC represents T_3; similarly ODC is a triangle of forces for the particle w_3, in which DC represents w_3, CO represents T_3 and OD represents T_4. It is seen at once that the points A, B, C, D lie in a vertical line and that the perpendicular ON on this line represents the horizontal component of the tension.

When the weights are equal we have $AB = BC = CD = \dots$ and the cotangent relation is obvious.

6·52. Chain of Heavy Rods. Let a series of uniform rods be smoothly jointed at their extremities so as to form a continuous chain hanging in a vertical plane with its ends fixed. Let $AB, BC, CD, \ldots$ represent consecutive rods of weights $W_1, W_2, W_3, \ldots$. The form of the chain would not be altered by supposing half of the weight of each rod to act at each of its ends. The weights would then act at the points $A, B, C, \ldots$ and the rods would play the part of the portions of string connecting the heavy particles in **6·5** and the form would be determined by the method of **6·5** or **6·51** with the difference that instead of the weight of a particle we have the mean of the weights of the two rods that meet in the corresponding point.

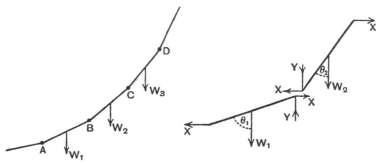

If, however, we want to determine the reactions at the joints we must replace the weights at the centres of the rods; then we may proceed as follows.

Let X, Y be the horizontal and vertical components of the reaction between the rods of weights W_1 and W_2, and let θ_1, θ_2 be the inclinations of these rods to the vertical. By taking moments for the lower rod about its lower end and for the upper rod about its upper end, we get

$$Y \sin \theta_1 - \tfrac{1}{2} W_1 \sin \theta_1 = X \cos \theta_1,$$

and
$$Y \sin \theta_2 + \tfrac{1}{2} W_2 \sin \theta_2 = X \cos \theta_2,$$

where the equations of moments have been divided by the lengths of the rods. Solving for X and Y, we get

$$X (\cot \theta_2 - \cot \theta_1) = \tfrac{1}{2} (W_1 + W_2) \quad \ldots\ldots\ldots\ldots\ldots(1),$$

and
$$Y (\tan \theta_1 - \tan \theta_2) = \tfrac{1}{2} (W_1 \tan \theta_1 + W_2 \tan \theta_2) \ldots\ldots\ldots(2).$$

It is evident that the horizontal component of reaction X is the same at every joint, and the result (1) is precisely the cotangent equation of **6·5** if the weight at the joint be assumed to be $\tfrac{1}{2}(W_1 + W_2)$ and X denotes the horizontal tension.

Again the X and Y at each joint compound into a single force R, and each rod is in equilibrium under the action of its weight and the reactions at its ends. The reactions at the ends of the rods must therefore intersect in pairs on the verticals through the middle points of the rods. The lines of the reactions there- fore form the sides of another polygon whose vertices are on the lines of action of the weights of the rods. Thus if AB, BC, CD, ... are con- secutive rods, $abcd$... is the sub- sidiary polygon along the sides of which act the reactions at the joints A, B, C, ...; and these forces along the sides of the polygon $abcd$..., if regarded as acting at the corners $b, c, d, ...$, i.e. at points on the lines of action of the weights $W_1, W_2, W_3, ...$, play the same part as the tensions of the strings in **6·5**.

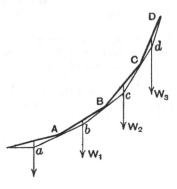

6·53. Example. *A chain of four equal heavy rods, each of weight w, is held up at points A, B in the same horizontal line. If the points of suspension are drawn apart until the horizontal component of the pulls at A, B are each $2w$, shew that $AB = 3\cdot54$ times the length of a rod and determine the slope of each rod.* [T.]

Since we are not concerned with the reactions at the joints but only with the form of the chain, we can suppose the weight of each rod to be equally divided between its ends.

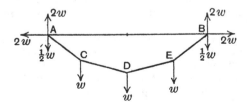

The vertical supporting forces at A and B must be equal by sym- metry, so that each is half the total weight, i.e. $2w$. The distribution of the weights at the ends of the rods is as shewn in the figure. These separate weights are now regarded as supported by tensions in the rods, and the horizontal component of tension is the same throughout and equal to $2w$, as we find by resolving horizontally at A or B.

Now construct a triangle of forces for the forces at the lowest joint D. Ocd is such a triangle in which cd represents w and Of the perpen- dicular from O to cd represents the horizontal component of tension $2w$. By symmetry Oc and Od parallel to DE and CD respectively are equally inclined to the horizontal so that $cf = fd = \frac{1}{2}w$. If we then draw a vertical bc equal to w, Obc will be a triangle of forces for the forces acting at E. The inclinations of DE and EB to the horizontal are the angles cOf and bOf or $\tan^{-1}\frac{1}{4}$ and $\tan^{-1}\frac{3}{4}$ respectively, i.e. 14° 3′ and 36° 54′.

Then AB being the projection of the four rods on the horizontal, we have

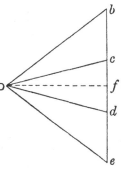

$$\frac{AB}{\text{length of a rod}} = 2\cos 14° \, 3' + 2\cos 36° \, 54'$$
$$= 3·54.$$

Alternatively we may solve the problem directly. Let α, β denote the inclinations of the lower and upper pairs of rods to the horizontal. By symmetry the reaction at D is horizontal, and by resolving horizontally for each rod we see that this reaction must be equal to the horizontal reaction at A, i.e. $2w$. Insert equal and opposite horizontal forces $2w$ at the ends of each rod. Then by resolving vertically for the rod CD, we see that the vertical reaction at C is w upwards on CD and downwards on CA.

Now take moments about C for the rod CD and we get

$2w \,.\, CD \sin\alpha = w \,.\, \frac{1}{2} CD \cos\alpha,$

or $\tan\alpha = \frac{1}{4}.$

And, by taking moments about A for the rod AC,

     $2w \,.\, AC \sin\beta = w \,.\, AC \cos\beta$
           $+ \, w \,.\, \frac{1}{2} AC \cos\beta,$

or $\tan\beta = \frac{3}{4}.$

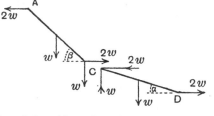

The remainder of the solution follows as before.

By comparing the two solutions it is evident that though the artifice of replacing the weight of a rod by equal weights at its ends may seem a simplification, it does not in point of fact reduce the amount of work necessary to the solution of such a problem as this one.

EXAMPLES

1. A sphere, of weight W, is in equilibrium on a smooth plane of inclination α to the horizontal, being supported by a string, which is of length equal to the radius and is fastened to two points, one on the sphere and one on the plane. Prove that the tension of the string is $\frac{1}{3}\sqrt{3}W\sin\alpha$.   [I.]

2. A heavy uniform rod AB rests with one end A in contact with a smooth vertical wall and with a string ACB attached to its ends and passing through a ring C vertically above A. Prove that

$$\tan BAC = 2\cot\tfrac{1}{2}(ACB). \qquad\qquad \text{[S.]}$$

3. A sphere, of weight W' and radius r, is hung by a string of length l from a point, and a uniform rod, of weight W and length $2a$, has one end attached to the same point and can freely turn about it. If the rod rests touching the sphere, shew that the angle θ made by the string with the vertical is given by the equation

$$\tan \theta = \frac{Wa\cos^2 \alpha}{W'r + Wa\sin \alpha \cos \alpha},$$

where $\cos \alpha = r/(l+r)$.

Find the tension of the string in terms of W, W', r, a, and α.   [S.]

4. A rod is in equilibrium resting over the rim of a smooth hemispherical bowl, fixed with rim horizontal, one end of the rod resting on the curved surface. Shew that the inclination, θ, of the rod to the horizontal is given by $4r\cos 2\theta = l\cos \theta$, where l is the length of the rod, and r the radius of the bowl. Shew also that for the rod to rest in this manner l must be greater than $\dfrac{2\sqrt{6}}{3}r$.   [I.]

5. A triangular lamina ABC, right-angled at A, rests with its plane vertical, and with the sides AB, AC supported by smooth fixed pegs D, E in a horizontal line. Prove that the inclination θ of AC to the horizontal is given by

$$AC\cos \theta - AB\sin \theta = 3DE\cos 2\theta.$$ [S.]

6. Two equal rods AB, AC hinged at A rest with their ends B, C on a smooth table and their plane vertical. A string is attached at points L, M of AB, AC respectively. Shew that the tension of the string is $W(BC)/4p$, where p is the perpendicular from A on LM and W is the weight of each rod.   [S.]

7. Two equal rods, AB, AC freely jointed at A, have their ends B, C connected by a light string. B and C rest symmetrically on two smooth planes inclined at angles α to the horizon, with their slopes facing one another, the plane of the rods being vertical. Shew that the tension of the string is $\frac{1}{2}W(\tan \theta - 2\tan \alpha)$, W being the weight of a rod and 2θ the angle BAC.   [S.]

8. Two uniform equal rods are hinged together and placed in a vertical plane, with their extremities on a smooth horizontal plane. They are kept from sliding by a horizontal rod the ends of which slide without friction on the other rods. Find the position of equilibrium and shew that the horizontal rod will rest close to the ground if its weight $= W\tan^2 \theta$, where W is the weight of either of the other rods, and θ their inclination to the vertical.   [S.]

9. Two uniform rods, each of weight W and length a, are freely jointed at A, and each passes over a smooth peg at the same level.

From A a weight W' is suspended. Shew that in the position of equilibrium the inclination θ of the rods to the horizon is given by

$$\cos^3 \theta = c(2W + W')/2Wa,$$

c being the distance between the pegs. [S.]

10. A circular disc of radius r and weight w is placed inside a smooth sphere of radius $R (> r)$, and at a distance d from the centre of the disc is fixed a weight W. Shew that in the position of equilibrium the plane of the disc makes with the horizon an angle whose tangent is

$$Wd/(W + w)\sqrt{R^2 - r^2}. \qquad \text{[S.]}$$

11. Three equal smooth pencils of weight W are tied together by a string and laid on a smooth table. Find the tension of the string when there is a pressure P between the pencils in contact with the table.

[S.]

12. Two spheres each of radius a and weight W lie in contact in a spherical bowl of radius na: shew that in the absence of friction the pressure between them is $W/\sqrt{n^2 - 2n}$. [S.]

13. A square lamina rests in a vertical plane on two smooth pegs at a distance c apart in the same horizontal plane. If θ be the angle the diagonal through the lowest corner makes with the vertical, x and y the distances of the centre of gravity from the sides which touch the pegs, prove that

$$(x + y)\sin \theta = (x \sim y)\cos \theta + \sqrt{2}c \sin 2\theta. \qquad \text{[S.]}$$

14. Three smooth heavy cylinders A, B, C lie on a table, with B between A and C and touching each of them. A and C have equal radii a and B has weight W and radius $b (< a)$. The cylinders A, C are pressed towards each other by equal forces P acting horizontally through their axes. Find the pressure of B on the table, if $b > a/4$ and the outer cylinders do not lift. [S.]

15. Three cylinders rest in equilibrium each touching a table so that two of the same radius a are in contact with the third of smaller radius, the two outer cylinders being pressed towards each other by equal forces P acting in horizontal lines at a distance a from the table. If the radius of the smaller cylinder be such that its pressure on the table is the greatest possible, then this pressure is $\frac{3}{2}P + W$, where W is the weight of the smaller cylinder. [I.]

16. An equilateral triangle, of side a, is supported by a string which has its extremities fastened to two angular points and passes through two small smooth rings fixed at a distance c from each other in a horizontal line. Prove that if the triangle hangs with one side vertical, the length of the string is $\sqrt{3}a - c$. Shew also that c cannot be greater than $a/\sqrt{3}$. [I.]

17. A string of length $2l$ has one end attached to the extremity of a smooth uniform heavy rod of length $2a$, and the other carries a weightless ring which slides on the rod. The rod is suspended by means of the string from a smooth peg: prove that if θ be the angle which the rod makes with the vertical, then $l\cos\theta = a\sin^3\theta$.   [S.]

18. A uniform smooth plank of weight W and length $2a$ is hinged to the bottom horizontal edge of a smooth plane, which is fixed at an inclination α to the horizontal, and a sphere of radius b and weight W' is placed between the plank and the plane; prove that, in the position of equilibrium the angle, θ, between the plank and the plane is given by the equation $2aW\sin^2\tfrac{1}{2}\theta\cos(\theta+\alpha) = bW'\sin\alpha,$

and the direction of the action at the hinge is inclined to the plank at an angle whose tangent is

$$(b - a\tan\tfrac{1}{2}\theta)\cot(\theta+\alpha)/b.   \text{[S.]}$$

19. A ladder of length $2a$ and weight W, with its centre of gravity three-eighths of the way up it, stands on a smooth horizontal plane resting against a smooth vertical wall, and the middle point is tied to a point in the wall by a horizontal rope of length l; find the tension of the rope.   [S.]

20. A ladder AB rests against a smooth vertical wall at A and is kept from slipping down by a cord PQ, fastened at Q to the ladder and at P to the line of intersection of the ground and the wall. Prove that the reactions at A, B and the tension in PQ are given by the following construction: Let the reactions at A, B meet in I, and let the vertical through the centre of gravity of the ladder meet AI in G and PQ in H: also let PQ meet BI in K. Then if the weight is represented by IK, the other sides of the quadrilateral $GHIK$ represent the reactions and tension.   [S.]

21. A uniform triangular lamina ABC rests between two smooth pegs D and E in the same horizontal, its vertex A being downwards. Shew that, if ϕ be the angle which the bisector of the angle at A makes with the vertical in the position of equilibrium,

$$\frac{3DE}{BC} = \frac{\sin\tfrac{1}{2}(B-C)}{\sin\phi}\sin^2\frac{A}{2} + \frac{\cos\tfrac{1}{2}(B-C)}{\cos\phi}\cos^2\frac{A}{2}.   \text{[C.]}$$

22. Two circular cylinders, of radii $3a$ and a, are glued together along a generating line and the compound body thus formed rests with the larger cylinder on a horizontal plane and both cylinders in contact with a third, of radius a, which itself rests on the plane. If all the bodies are smooth and the generating lines are all parallel and horizontal and the weight of the upper small cylinder is W, prove that the total pressure between the two lower cylinders will be $\dfrac{15-\sqrt{5}}{40}W$.   [S.]

23. A uniform rod, of length c, rests with one end on a smooth elliptic arc whose major axis is horizontal and with the other on a smooth vertical plane at a distance h from the centre of the ellipse; the ellipse and the rod both being in a vertical plane. Prove that, if θ is the angle which the rod makes with the horizontal, and $2a$, $2b$ are the axes of the ellipse,
$$2b \tan \theta = a \tan \phi,$$
where
$$a \cos \phi + h = c \cos \theta;$$
and explain the result when
$$a = 2b = c, \quad h = 0. \qquad \text{[S.]}$$

24. A uniform rod of length $2a$ and weight W hangs in an oblique position, supported by a light inextensible string of length $2l \, (l > a)$ whose ends are fastened to the ends of the rod, and which passes over a smooth peg, and a weight w is attached to the rod at a distance d from its middle point. Prove that the lengths of the string on the two sides of the peg are
$$l\left(1 - \frac{d}{an}\right) \quad \text{and} \quad l\left(1 + \frac{d}{an}\right),$$
where $nw = W + w$. \qquad [S.]

25. AB, BC are two uniform heavy rods of equal length and weight W. The rod AB can move freely about A, there is a hinge at B, and at C there is a ring which can move freely along a fixed rod through A which is inclined downwards at an angle α to the horizontal. Shew that in equilibrium $\tan BAC = \frac{1}{2} \cot \alpha$, and that the horizontal component of the action at B is $\frac{3}{8} W \sin 2\alpha$. \qquad [S.]

26. Two uniform rods AB and CD each of weight W and length a are smoothly jointed together at a point O, where OB and OD are each of length b. The rods rest in a vertical plane with the ends A and C on a smooth table and the ends B and D connected by a light string. Prove that the reaction at the joint is $\dfrac{aW}{2b} \tan \alpha$, where α is the inclination of either rod to the vertical. \qquad [S.]

27. A uniform rod of length $2l$ is attached by smooth hinges at its ends to a vertical wall, and to a uniform smooth sphere of the same weight as the rod, which rests against the wall. Prove that the co-tangents of the angles which the rod, and the actions at the hinges, make with the vertical are proportional to $3 : 2 : 4$. Prove also that, if the rod is inclined to the vertical at an angle $\cot^{-1} 2$, the radius of the sphere is $\sqrt{5}\, l$. \qquad [S.]

28. Two light rods AO, OB are jointed at O. The rod OA is attached by a hinge at A to a fixed rod AD and B is attached to a ring which can slide along AD. A force P acts at O towards AB at right angles to AB and a force Q acts at B in the direction BA. The joint, hinge and ring are frictionless and the angles OAB, OBA are acute; prove that in equilibrium
$$P \cos OAB \cos OBA = Q \sin AOB. \qquad \text{[S.]}$$

29. Two equal beams AB, AC each of weight W connected by a hinge at A are placed in a vertical plane with their extremities B, C resting on a horizontal plane; they are kept from falling by strings connecting B and C with the middle points of the opposite sides; shew that the tension of either string is

$$\frac{W}{8}\sqrt{1+9\cot^2\theta},$$

where θ is the inclination of either beam to the horizon. Shew also that the action of the hinge on either rod is $\frac{3}{4}W\cot\theta$. [S.]

30. Two equal uniform rods AB, AC, each of weight W, are smoothly jointed at A: D, E are points on AB, AC respectively, such that $AD = AE = \lambda$. AB and two light equal strings connect B to E and C to D. The system stands in a vertical plane on a smooth horizontal table. Shew that the tension in either string is represented by BE or CD, provided that W is represented by 8λ times the line joining A to the mid-point of BC. [S.]

31. Two equal rods of weight w are freely jointed and their free ends are attached by strings to a fixed point. A circular disc of weight W and radius r rests in the angle between the rods, and the whole hangs in a vertical plane. If $2a$ be the length of each rod and each string, and 2θ the angle between the rods, prove that

$$r = 2a\sin^2\theta\tan\theta\frac{3w+2W}{W}.$$ [S.]

32. Three rods of the same material are freely jointed so as to form a triangle ABC. The middle point of AB is supported; shew that, when the triangle is in equilibrium, AB makes an angle with the horizon $\frac{1}{2}(A-B)$ and that the action on either rod at the joint C makes the same angle with the horizon and is of magnitude

$$\tfrac{1}{4}\{(W'-W)^2+(W'+W)^2\cot^2\tfrac{1}{2}(A+B)\}^{\frac{1}{2}};$$

W, W' being the weights of the rods AC, BC. [S.]

33. An elliptic wire is fixed with major axis vertical. Two smooth rings slide on it, being connected by a string passing through a smooth ring at the upper focus. Shew that if the weights of the rings are equal equilibrium is possible in all positions, but if they are unequal one ring must be at the highest or the lowest point, according as the length of the string is less or greater than the major axis. [S.]

34. A hollow right circular cylinder with smooth interior and open at both ends is placed on a horizontal plane. A series of n uniform rods of equal weight (w) and of equal length (greater than the diameter of the cylinder) are placed one on the top of another and resting against the inside of the cylinder. Shew that the cylinder will not fall over, if its weight is greater than nw. [I.]

35. A triangle ABC hung up by the corner A is formed of uniform rods cut from the same uniform bar and smoothly jointed at A, B, C. The rod AB is loaded so that BC is horizontal. Find the action on the rod AB at B, and prove that its direction is inclined to the vertical at an angle whose tangent is

$$\cot C\left(1+\frac{\sin B}{\sin A}\right).\qquad\text{[S.]}$$

36. One end of a uniform rod is hinged to a fixed point, a cord attached to the other end passing over a pulley sustains a weight, the rod and pulley being in the same vertical plane. If the system be in equilibrium with the rod inclined at a given angle i to the horizontal, and the weight attached to the cord is the least possible, shew that the pulley lies on a fixed straight line, and find the position of this line. Also find the ratio of the attached weight to the weight of the rod. [S.]

37. A smooth lamina in the form of an equilateral triangle ABC is placed on a horizontal table so that each of its sides touches one of three smooth pegs P, Q, R, where

$$AR:RB=BP:PC=CQ:QA=n:1.$$

A couple G acts on the lamina; shew that the pressure on each of the pegs is $\dfrac{2}{3}\cdot\dfrac{n+1}{n-1}\cdot\dfrac{G}{BC}.$ [S.]

38. A triangle ABC is formed of three light rods smoothly jointed to each other at their ends. D is the foot of the perpendicular from A on BC. A string at tension T joins AD. Find the stresses in the rods AB and AC. [S.]

39. $ABCD$ is a square formed of four light rods jointed together, the diagonal AC being a fifth light rod. Weights P and Q are attached to the corners B, D respectively, and the system is hung up by the corner A. Find the inclination to the vertical of the rod AC and also the stress in it. [S.]

40. Four equal rods hinged together form a rhombus $ABCD$, which is hung in a vertical plane from a fixed point E by two equal rods EB and ED and a third EA, of such lengths that the angle at C is 60° and each of the angles at E is 15°; the rods being without weight and a weight W attached to C, prove that the force in the rod EA is

$$W(\sqrt{3}+1).\qquad\text{[S.]}$$

41. $ABCD$ is a rhombus formed of freely jointed light rods. AC is vertical, A being the higher end, and B, D are tied by strings of equal length to a fixed point in the line AC. Weights W, W' ($W>W'$) are suspended from A, C. If the rhombus is constrained to remain in a vertical plane, prove that in the position of equilibrium the fixed point divides AC in the ratio $W':W$. [S.]

42. A rhombus $ABCD$ of smoothly jointed rods rests on a smooth table with the rod BC fixed in position. The middle points of AD, DC are connected by a string which is kept taut by a couple L applied to the rod AB. Prove that the tension of the string is

$$2L/(AB)\cos(\tfrac{1}{2}ABC). \qquad\qquad [\text{S.}]$$

43. Four light rods AB, BC, CD, DA are freely jointed together; $AB = BC$ and $CD = DA$. The rod AB is fixed horizontally and masses P, Q are suspended from C, D respectively. Shew that in equilibrium the angles DAB, ABC will be both acute or both obtuse, and that if α, β are the angles which AD, CD make with the vertical

$$\frac{\sin 2\beta}{\sin 2\alpha} = 1 + \frac{Q}{P}. \qquad\qquad [\text{S.}]$$

44. Five weightless rods form a regular pentagon $ABCDE$, and the framework is stiffened by other weightless rods BD, CE. The system is placed in a vertical plane with CD on a horizontal table, and a weight W is hung from A. Prove that the thrust in BD or CE is $W \cot \tfrac{1}{5}\pi$.

$$[\text{S.}]$$

45. Two corners of a regular pentagon of light freely jointed rods are connected by a string in tension, and equilibrium is maintained by another string also in tension, connecting one corner to the middle point of the opposite side, the strings being perpendicular to each other. Shew that their tensions are in the ratio $2\sin\tfrac{1}{5}\pi : 1$. $\qquad [\text{S.}]$

46. Six equal weightless rods are smoothly jointed at their ends so as to form a regular tetrahedron $ABCD$. Prove that if equal weights W are attached to the joints C, D, while the rod AB is supported in a horizontal position, the thrust along the rod CD will be $W/\sqrt{2}$. $\quad [\text{S.}]$

47. Six equal rigid weightless bars, freely jointed at their ends, form a regular tetrahedron $ABCD$. It is suspended from A, and three weights each equal to W are hung from B, C and D respectively. Find the stresses in all the bars. $\qquad\qquad [\text{S.}]$

48. A framework in the form of a tetrahedron is formed of six equal light rods freely jointed to each other at their ends. A string at tension T joins the middle points of two opposite rods. Prove that the stress in each of the other rods is $T/2\sqrt{2}$. $\qquad\qquad [\text{S.}]$

49. A smoothly jointed quadrilateral of rods lies on a smooth horizontal table and is enclosed in a smooth circular hoop which presses tightly at each of the hinges. Prove that the pressures at the hinges are proportional to the sides of the circumscribed quadrilateral which touches the hoop at the hinges and that the stresses in the rods are inversely proportional to their distances from the centre of the hoop.

$$[\text{S.}]$$

50. AB represents the piston-rod of the fixed cylinder of a steam-engine, and CD is a crank turning an axle D, BC being a connecting-rod. DE is drawn perpendicular to AB meeting BC in E, and CF is the perpendicular from C on AB; CG is perpendicular to BC.

Shew that if the thrust in AB is given,

    (i) the couple exerted on the axle is proportional to DE;

    (ii) the transverse pressure on the guides constraining the joint B to move along AD is proportional to CG;

    (iii) the thrust in BC is inversely proportional to BF.   [S.]

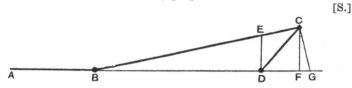

51. Four equal uniform rods are freely jointed together so as to form a square $ABCD$, and the system hangs from the point A, the square form being maintained by an inextensible string connecting the middle points of AB and BC. If W is the weight of each rod, prove that the tension of the string is $4W$, that the stress at B is $\frac{1}{2}W\sqrt{17}$ in the direction inclined to the vertical at the angle $\tan^{-1}\frac{1}{4}$, that the stress at C is $\frac{1}{2}W\sqrt{5}$ in the direction inclined to the vertical at the angle $\cot^{-1}2$, and that the stress at D is $\frac{1}{2}W$ in a horizontal direction.   [S.]

52. A framework of four heavy rods, of length a, hinged together to form a rhombus is supported by a smooth cylinder of radius c. Shew that, if the rods are in equilibrium when each makes an angle $30°$ with the vertical, then $a = 4\sqrt{3}c$. Find the ratio of the actions at the top and bottom hinges.   [S.]

53. Four equal uniform straight rods are freely jointed together to form a rhombus $ABCD$, the rod AD being fixed vertically with A uppermost. A string CE joins the corner C to a point E in AB; prove that the tension of the string bears to twice the weight of a rod the ratio of CE to BE. Find also the magnitude and direction of the stress at the joint B.   [S.]

54. Four heavy uniform rods AB, BC, CD, DA, of which $AB = AD$ and $BC = CD$, are jointed at their extremities, and the corners A and C joined by a string of such a length that AB and BC are at right angles. If the framework is suspended from A, prove that the tension of the string is

$$wl\sin\alpha\,(1 + \sin\alpha\cos\alpha + \sin^2\alpha),$$

where l is the length of the string, α the inclination of the upper rods to the vertical, and w is the weight of the rods per unit length.   [S.]

55. A pentagon $ABCDE$ is formed of rods whose weight is w per unit length. The rods are freely jointed together and stand in a vertical plane with the lowest rod AB fixed horizontally, while the joints C, E are connected by a string. If BC and AE are of length a, CD and DE of length b and the angles at A and B are each 120° and the angles at C and E are each 90°, shew that the tension of the string is

$$w\frac{a+5b}{2\sqrt{3}}.$$ [S.]

56. Five equal uniform rods, each of weight W, are freely jointed at their extremities, so as to form a pentagon. The pentagon is suspended in a vertical plane by a string attached to one corner, and the two adjacent corners are connected by a light rod of such length that the pentagon is regular. Shew that the stress in this rod is

$$W(\tan 18° + 2\tan 54°),$$

and find the reactions at the corners of the pentagon. [S.]

57. A regular hexagon $ABCDEF$ is made of six equal uniform rods jointed freely. The hexagon rests in a vertical plane having AB in contact with a given horizontal plane and C, F are connected by a light inextensible string; shew that the tension of the string is $W\sqrt{3}$, where W is the weight of a rod, and shew that the action at the joint E is $\frac{W}{2}\sqrt{\frac{7}{3}}.$ [S.]

ANSWERS

3. $W(Wa\cot\alpha + W'r)/\sqrt{(W^2a^2\cos^2\alpha + 2WW'ar\sin\alpha\cos\alpha + W^2r^2)}$.

11. $P + W/2\sqrt{3}$.   14. $W + P(a-b)/\sqrt{(ab)}$.   19. $3Wl/4\sqrt{(a^2-l^2)}$.

35. $\frac{1}{2}\sqrt{\{w^2 + (w+w')^2\cot^2 C\}}$, where w, w' are the weights of BC, CA.

36. Perpendicular to the rod through its free end. $\frac{1}{2}\cos i$.

38. $T\csc A\cos C$, $T\csc A\cos B$.

39. $\tan^{-1}\{(P-Q)/(P+Q)\}$, $\sqrt{2PQ}/\sqrt{(P^2+Q^2)}$.

47. $W/\sqrt{6}$, $W\sqrt{(3/2)}$.   52. 11 : 1.

53. $w\left(\frac{1}{2} + \frac{2AE}{BE}\right)$ vertically, where w = wt. of a rod.

Chapter VII

BENDING MOMENTS

7·1. In the preceding chapter we considered a number of problems about the equilibrium of rods or beams, but beyond an occasional reference to the fact that a light rod may be regarded as a 'tie' or a 'strut' exerting equal and opposite forces at its ends along its length, no reference has yet been made to the sort of stresses that are set up in a rod or beam by externally applied forces. We shall discuss this subject in the present chapter.

7·2. Let AB be a beam in equilibrium under the action of any given system of external forces. Imagine the beam to be divided into two parts AC, CB by a cross section at any point C; and consider the equilibrium of one part of the beam, say

CB. We may assume that some, but not all, of the given external forces act on the part CB, and that these forces which act upon CB would not in general be in equilibrium of themselves, because the other forces which act upon AC are needed to balance them. But the part CB of the beam *is* in equilibrium, therefore the external forces which act upon it must be balanced by stresses at C exerted by the part AC upon the part CB.

It follows that the stresses in the beam at any cross section C are a system of forces that would balance the external force system which acts upon either portion into which the beam is divided at C.

Now confining ourselves to coplanar forces, we saw in **5·1** that any system of coplanar forces can be reduced to a single force acting at an arbitrarily chosen point together with a couple. Hence the external forces acting on the part CB are equivalent to a single force acting at C together with a couple.

Let the force be resolved into components T along the beam and S at right angles to it, and let M be moment of the couple. Then T is called the tension or **thrust** according as it is a force tending to lengthen or shorten the beam; S is called the shearing force, and M is called the **bending moment** at C.

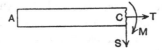

This diagram shows the external forces on CB and on AC respectively each reduced to act at the point C and constituting equal and opposite systems.

Now the forces exerted by AC on CB must balance the external forces on CB, so that we may represent the forces exerted by AC on CB thus:

and the equal and opposite forces exerted by CB on AC thus:

By reference to **5·1** we see that the shearing force and the tension or thrust may be found by moving all the external forces that act upon CB (or AC) parallel to themselves to act at C, and that the bending moment at C is the algebraical sum of the moments about C of all the external forces which act upon CB (or AC). The sense of the shearing force and of the bending moment are determined in this way.

7·21. Examples. (i) *A light beam of length l is supported at its ends and carries a load W concentrated at a point at a distance a from one end.*
Let $AB = l$ be the beam, and let $AC = a$.
By taking moments about B, we find that the supporting force at A is $W\left(1 - \dfrac{a}{l}\right)$, and therefore that at B is $W\dfrac{a}{l}$.

Let P be any point on the beam at a distance x from A, then when P lies between A and C,
or for $0 < x < a$, we have

$$S = W\left(1 - \frac{a}{l}\right) \text{ upwards,}$$

and　$M = W\left(1 - \frac{a}{l}\right)x$ clockwise;

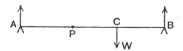

and when P lies between C and B,
or for $a < x < l$,

S is the resultant of $W\left(1 - \frac{a}{l}\right)$ upwards and W downwards,

i.e.　　　　　　　　　$S = W\frac{a}{l}$ downwards,

and M is the sum of the moments of the same two forces,

i.e.　　$M = W\left(1 - \frac{a}{l}\right)x - W(x - a) = Wa\left(1 - \frac{x}{l}\right)$ clockwise.

These values of S and M have been found from the forces on AP; if we used instead the forces on PB we should clearly get the same numerical values but acting in the opposite sense.

Graphical representations of the values of the shearing force S and the bending moment M are useful as shewing at a glance the way in which they vary along the beam.

Ordinates are drawn at points along the beam to represent S or M as the case may be. In this example the graph of S is two lines ac, $c'b$ parallel to the beam with a jump in the value in passing the load W; and the graph of M is two straight lines which pass through A and B and meet on the vertical through C, shewing no discontinuity in M in passing

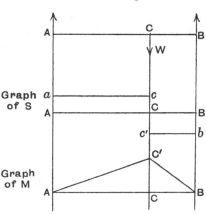

the load W, though there is a discontinuity in the gradient of M.

(ii) *A heavy uniform beam of length l supported at one end and at a distance $\frac{1}{3}l$ from the other.*

Let ABC be the beam supported at B and C. Let w denote the weight of unit length, so that wl is the total weight. Then by taking moments

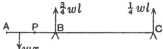

R S　　　　　　　　　　　　　　　　　　　　　　　　　　7

about C we find that the supporting force at B is $\frac{3}{4}wl$, so that that at C is $\frac{1}{4}wl$.

Let x denote the distance from the end A of a variable point P. We propose to find the values of the shearing force S and the bending moment M at P as P moves along the beam from A to C.

So long as P lies between A and B, i.e. for $0 < x < \frac{1}{2}l$, the only force on AP is its weight wx acting at a distance $\frac{1}{2}x$ from P, so that

$$S = wx \text{ downwards,}$$

and $\qquad\qquad M = \frac{1}{2}wx^2 \text{ counter-clockwise.}$

As P moves to the right of B we have to include in the forces acting upon AP the upward force $\frac{3}{4}wl$ at B as well as the weight wx, so that for $\frac{1}{2}l < x < l$

$$S = wx - \tfrac{3}{4}wl \text{ downwards,}$$

and $\ M = \frac{1}{2}wx^2 - \frac{3}{4}wl(x - \frac{1}{4}l)$
$\qquad = \frac{1}{4}w(l-x)(l-2x)$
$\qquad\qquad$ counter-clockwise.

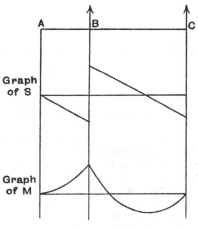

Graph of S

Graph of M

If we draw the graphs of S and M, we notice that the graph of S consists of parts of two parallel straight lines, with a discontinuity at B; that the graph of M consists of arcs of two parabolas, and that M vanishes at the ends and the middle of the beam and has no point of discontinuity in its values though there is a discontinuity in the gradient at the point B.

7·3. The examples just considered shew that the shearing force and the gradient of the bending moment are liable to discontinuities. We will now formulate some general relations between shearing force and bending moment for a straight beam and see why and when such discontinuities are to be expected to arise.

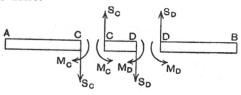

Let us disregard tension and thrust and suppose that any

length AC ($=x$) of a beam is acted upon by external forces which give rise to a shearing force S_C and a bending moment M_C at C. As it is necessary to adopt some convention about signs, we will suppose that S_C acts downwards and M_C in the clockwise sense on the portion of the beam on the left of C, and of course in the contrary senses on the portion of the beam on the right of C, and that a like convention is adopted all along the beam.

Consider a short length CD of the beam. The shearing force and bending moment at C are denoted by S_C and M_C, and in like manner those at D are denoted by S_D and M_D acting in the senses indicated in the figure in accordance with our convention. We may consider three cases:

(i) *When CD is subject to no forces but shearing force and bending moment:* by resolving vertically,

$$S_D = S_C \qquad \dots\dots\dots\dots\dots\dots\dots(1),$$

and by taking moments about D,

$$M_D - M_C + S_C \cdot CD = 0,$$

which, as the length CD (or dx) tends to zero, gives a differentia relation

$$\frac{dM}{dx} = -S \qquad \dots\dots\dots\dots\dots\dots\dots(2).$$

(1) shews that in this case the shearing force is constant, and (2) shews that at all points of CD the gradient of the bending moment is equal to minus the shearing force.

(ii) *When CD carries a uniformly distributed load, such as its weight:* by resolving and taking moments as above, we get

$$S_D - S_C + w \cdot CD = 0$$

and $\qquad M_D - M_C + S_C \cdot CD - \tfrac{1}{2}w \cdot (CD)^2 = 0,$

where w is the load per unit length.

These, as the length of CD diminishes, give differential relations

$$\frac{dS}{dx} = -w \qquad \dots\dots\dots\dots\dots\dots\dots(3),$$

and $\qquad\qquad \dfrac{dM}{dx} = -S \qquad \dots\dots\dots\dots\dots\dots\dots(4),$

so that the gradient of the graph of the shearing force is minus the load per unit length and is constant for a uniformly distributed load. We also see from (4) that the gradient of the bending moment is equal to minus the shearing force whether the beam is loaded in this manner or not, and that a discontinuity in the shearing force implies a discontinuity in the gradient of the bending moment.

These facts are exemplified in the examples of 7·21; as also the fact that the vanishing of the shearing force is accompanied by the vanishing of the gradient of the bending moment.

(iii) *When CD carries a finite load W concentrated at a point on CD:* by resolving and taking moments as above, we get

$$S_D - S_C = -W \quad \dots\dots\dots\dots\dots\dots(5),$$

and $$M_D - M_C + S_C \cdot CD - W \cdot \epsilon CD = 0 \dots\dots\dots\dots(6),$$

where $0 < \epsilon < 1$.

It appears from (5) that a concentrated load gives rise to a discontinuity in the shearing force, as is otherwise obvious; but if in (6) the points C and D move up to the point at which the load is concentrated so that $CD \to 0$, we get $M_D = M_C$, so that no discontinuity arises in the bending moment, but inasmuch as there is a discontinuity in the shearing force there will also be a discontinuity in the *gradient* of the bending moment.

7·4. Examples. (i) *A light rod ABCD, in which AB = 5 ft., BC = 10 ft. and CD = 5 ft., is supported at B and C and carries uniformly distributed loads of 1 lb. per foot along AB and CD and a load of 20 lb. concentrated at the middle point E of BC. Draw diagrams to represent the shearing force and bending moment at all points of the rod.*

Since there is symmetry about the centre and the total load is 30 lb., therefore the supporting forces at B and C are each 15 lb. If x denotes the distance from A of a variable point P which moves across the beam from left to right, then the shearing force and bending moment at this point are given by the following set of values:

when $0 < x < 5$, $\qquad\qquad S = x$ lb.,

$$M = \tfrac{1}{2}x^2 \text{ ft. lb.;}$$

$5 < x < 10,$ $\qquad S = 5 - 15 = -10$ lb.,

$\qquad\qquad\qquad M = 5\,(x - \tfrac{5}{2}) - 15\,(x - 5)$

$\qquad\qquad\qquad\quad = (-10x + 1\tfrac{25}{2})$ ft. lb.;

$10 < x < 15,$ $\qquad S = 5 - 15 + 20 = 10$ lb.,

$\qquad\qquad\qquad M = 5\,(x - \tfrac{5}{2}) - 15\,(x - 5) + 20\,(x - 10)$

$\qquad\qquad\qquad\quad = (10x - 2\tfrac{75}{2})$ ft. lb.;

$15 < x < 20,$ $\qquad S = 5 - 15 + 20 - 15 + (x - 15)$

$\qquad\qquad\qquad\quad = -(20 - x)$ lb.,

$\qquad\qquad\qquad M = 5\,(x - \tfrac{5}{2}) - 15\,(x - 5) + 20\,(x - 10)$

$\qquad\qquad\qquad\qquad\qquad - 15\,(x - 15) + \tfrac{1}{2}\,(x - 15)^2$

$\qquad\qquad\qquad\quad = \tfrac{1}{2}\,(20 - x)^2$ ft. lb.

It will be noticed that in this procedure so long as P lies between A and B, the effective load on AP is x lb., but when P is to the right of AB the whole load on AB contributes to the shearing force and bending moment; and that when P is to the right of C we have to take into account the load on CP, viz. $(x - 15)$ lb.

Also, when P lies on CD, S and M could easily be written down by considering the force on PD which is simply the load $(20 - x)$ lb.; we may use this as a check on the results obtained above, noting the difference of sign according as the force and couple are considered to be acting upon AP or upon PD.

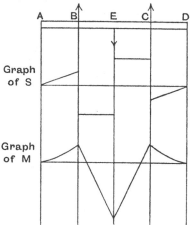

(ii) *A uniform rod of length $3l$ and weight w per unit length rests upon two supports at its points of trisection. A peg of small diameter is fixed horizontally through the centre of the rod and a couple of moment $\tfrac{1}{2}wl^2$ is applied to the axis of the peg. Sketch carefully diagrams of shearing force and bending moment.* [T.]

Let OD be the rod, A, B the points of trisection and C the middle point. Then the total weight is $3wl$, and if R and Q denote the supporting forces at A and B, we have $R + Q = 3wl$,

and by moments about B

$$Rl + \tfrac{1}{2}wl^2 = \tfrac{3}{2}wl;$$

whence $\qquad\qquad R = wl \quad \text{and} \quad Q = 2wl.$

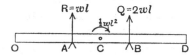

Then for the shearing force and bending moment at a point P at a distance x from O we have the following values:

when $0 < x < l$, $S = wx,$

$M = \tfrac{1}{2}wx^2;$

$l < x < \tfrac{3}{2}l,$ $S = wx - R = w(x - l),$

$M = \tfrac{1}{2}wx^2 - R(x - l)$

$= \tfrac{1}{2}w(x^2 - 2lx + 2l^2);$

$\tfrac{3}{2}l < x < 2l,$ $S = w(x - l),$

$M = \tfrac{1}{2}wx^2 - R(x - l) - \tfrac{1}{2}wl^2$

$= \tfrac{1}{2}w(x - l)^2;$

$2l < x < 3l,$ $S = wx - R - Q = -w(3l - x),$

$M = \tfrac{1}{2}wx^2 - R(x - l) - \tfrac{1}{2}wl^2 - S(x - 2l)$

$= \tfrac{1}{2}w(3l - x)^2.$

This is an example in which there is a discontinuity in the value of the bending moment. It arises from the fact that there is an externally applied couple at a particular point of the beam.

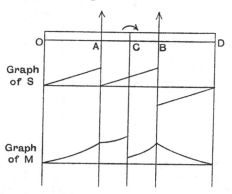

Graph of S

Graph of M

(iii) *A uniform beam of length $2l$ rests symmetrically on two supports which are a distance $2a$ apart in a horizontal line; prove that the beam is* *least liable to break if $a = l(2 - \sqrt{2})$, it being assumed that the beam is liable to break if a definite bending moment is exceeded at any point.* [S.]

Let AB be the beam, C its middle point and D, E the points of support, so that $AD = EB = l - a$. If w is the weight of unit length, the supporting forces are each wl.

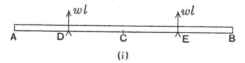

(i)

If x denotes the distance of a point P from A, then for the bending moment at P, we have that

when $0 \leqslant x \leqslant l - a$,

$M_1 = \frac{1}{2} w x^2$ with a greatest value $\frac{1}{2} w (l - a)^2$ at D.

And when $l - a \leqslant x \leqslant l + a$,

$$M_2 = \frac{1}{2} w x^2 - w l (x - l + a)$$
$$= \frac{1}{2} w \{(x - l)^2 - l (2a - l)\}.$$

Now if $l > 2a$, then M_2 cannot vanish and its values vary from a greatest value $\frac{1}{2} w (l - a)^2$ at D to a least value $\frac{1}{2} w (l^2 - 2al)$ at C.

The diagram of bending moments is as shewn in fig. (ii).

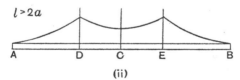

(ii)

In this case the beam must be able to stand a bending moment $\frac{1}{2} w (l - a)^2$, the value at D and E; this decreases as a increases, but since $a < \frac{1}{2} l$, it cannot be less than $\frac{1}{8} w l^2$, which is its value when $a = \frac{1}{2} l$.

Again if $l < 2a$, then M_2 is negative at C and the diagram of bending moments is as shewn in fig. (iii).

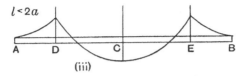

(iii)

The numerical values of M_2 at D and at C are then $\frac{1}{2} w (l - a)^2$ and $\frac{1}{2} w l (2a - l)$. The former is decreased and the latter is increased by increasing a. It follows that the value of a which will give the lowest maximum to M_2 is found by making the above two values equal, i.e. when

$$(l - a)^2 = 2al - l^2,$$

or when

$$a = l (2 - \sqrt{2}),$$

and this makes the greatest bending moment $\frac{1}{2}wl^2(\sqrt{2}-1)^2$ which is less than $\frac{1}{8}wl^2$, so that this value of a gives the lowest value to the maximum bending moment along the beam.

EXAMPLES

1. A uniform heavy rod AB is supported at each end. If w be the weight per unit length, prove that the bending moment at any point P is $\frac{1}{2}wAP.PB$.

2. A light horizontal rod 16 ft. long is supported at points distant 2 ft. from one end and 3 ft. from the other, and is loaded with a weight of 5 lb. at its middle point and weights of 3 lb. at each end. Draw diagrams shewing the shearing stress and bending moment at all points of the rod.   [T.]

3. A beam AD, 35 ft. long and supported at points B and C distant 10 and 5 ft. respectively from the ends A and D, carries a load of 5 tons at D. If the weight of the beam is neglected, draw diagrams shewing the shearing force and bending moment at all points of the beam. [T.]

4. A light rod is maintained in a horizontal position by supports at its ends. Find expressions for the bending moment as a concentrated load is moved slowly along the rod, and shew that the tendency to break is greatest at the load and has its maximum value when the load is in the middle of the rod.

5. A light rod is supported at A and B and carries a load at Q. Prove that the bending moment at P is proportional to $AP.QB$ or to $AQ.PB$ according as P lies on AQ or on QB.

6. A uniform rod ABC of length a is smoothly hinged at A and supported at B at an inclination of $45°$ to the horizontal, where $AB=\frac{3}{4}a$. Find expressions for the bending moment at all points of the rod and shew that it vanishes at a distance $\frac{2}{3}a$ from the end A.

7. A light rod passes through two smooth closely fitting fixed rings at its ends and is acted upon by equal and opposite forces at its points of trisection. Draw a diagram to shew the bending moment at every point of the rod.

8. Two uniform heavy rods AB, BC of the same weight per unit length are smoothly jointed at B and maintained in a horizontal position by supports at A, D and C, where $AD=3DB=3BC$. Find the pressures on the supports and draw diagrams shewing the shearing force and bending moment at all points of the rods.

9. Two equal uniform beams AB, BC of length a and of the same weight per unit length w are smoothly hinged at B and supported in a horizontal line by props at A and D, where $BD=\frac{1}{2}DC$. Find expressions for the shearing force and bending moment at any point of each beam and draw graphs to represent the variations in their values.

[S.]

10. A uniform beam AB of length l and weight w per unit length is smoothly hinged at A, and is kept at an inclination of $45°$ to the upward vertical through A by a light horizontal rope which joins a fixed point to the point of the beam distant $l/3$ from A. Find expressions for the thrust, shearing force and bending moment at any point of the beam.

[S.]

11. Two uniform rods AC, CB of the same weight per unit length are smoothly jointed at C, and smoothly hinged to two points A, B in the same vertical line. Prove that the reaction at C bisects the angle ACB, and that the bending moments at two points P, Q on AC and BC respectively are in the ratio

$$AP.PC\sin A \text{ to } BQ.QC\sin B.$$

12. A uniform horizontal beam, which is to carry a uniformly distributed load, is supported at one end and at some other point; find where the second support should be placed in order that the greatest possible load may be carried by the beam, and shew that it will divide the beam in the ratio 1 to $\sqrt{2}-1$. [T.]

13. A uniform semicircular hoop hangs freely from one of its ends. Prove that the bending moment is greatest at the point where the tangent is vertical.

14. A beam AB of uniform material projects horizontally from a wall at A. It is of length l and its cross section tapers uniformly from a^2 at A to b^2 at B. Prove that the greatest bending moment is

$$\tfrac{1}{12}wl^2(a^2+2ab+3b^2),$$

where w is the density of the material.

15. Two uniform rods ACB and CD are smoothly jointed at C the middle point of AB and the ends A, D are smoothly hinged to points in the same vertical line with D below A and such that $DA=\tfrac{1}{2}DC$. Determine the values of the tension or thrust, the shearing force and the bending moment in AB and CD,

(i) when the rods are light and a load W is suspended at B;

(ii) when the rods are of weight w per unit length.

ANSWERS

6. On CB, $\dfrac{1}{2\sqrt{2}}wx^2$; on BA, $\dfrac{w}{6\sqrt{2}}(3x-a)(x-a)$, x measured from C.

8. $\tfrac{7}{6}W$ at A, $\tfrac{10}{3}W$ at D, $\tfrac{1}{2}W$ at C, when $W=$ wt. of BC.

9. $0<x<\tfrac{1}{3}a$, $S=\tfrac{1}{2}w(a-2x)$, $M=\tfrac{1}{2}wx(a-x)$;
$\tfrac{1}{3}a<x<2a$, $S=w(2a-x)$, $M=-\tfrac{1}{2}w(2a-x)^2$, x measured from A.

10. $0<x<\tfrac{2}{3}l$, $T=S=wx/\sqrt{2}$, $M=wx^2/2\sqrt{2}$;
$\tfrac{2}{3}l<x<l$, $T=w(3l+2x)/2\sqrt{2}$, $S=-w(3l-2x)/2\sqrt{2}$,
$M=w(2l-x)(l-x)/2\sqrt{2}$, x measured from B.

15. (i) On CB, thrust

$$T = -W \cos DAB, \quad S = W \sin DAB,$$
$$M_P = W \cdot BP \cdot \sin DAB;$$

on AC, $\quad\quad T = -W \cos DAB - 4W \cos ACD,$

$$S = -W \sin DAB, \quad M_Q = W(BQ - 2CQ) \sin DAB;$$

on CD, $\quad\quad\quad\quad\quad\quad T = 4W, \quad S = 0, \quad M = 0.$

(ii) x is measured from B on ACB and from C on CD.

On CB, $\quad\quad T_x = -wx \cos DAB, \quad S_x = wx \sin DAB,$

$$M_x = \tfrac{1}{2} wx^2 \sin DAB;$$

on AC, $\quad\quad T_x = -wx \cos DAB - wa(1 + 4 \cos ACD),$

$$S_x = -w(2a - x)\sin DAB, \quad M_x = \tfrac{1}{2} w(2a - x)^2 \sin DAB;$$

on CD, $\quad\quad T_x = wx \cos ADC + wa(4 + \cos ACD),$

$$S_x = -w(b - x)\sin ADC, \quad M_x = -\tfrac{1}{2} wx(2b - x)\sin ADC,$$

where $AC = a$ and $AD = b$.

Chapter VIII

GRAPHICAL STATICS

8·1. *To determine graphically the magnitude, direction and position of the resultant of any number of coplanar forces.*

Let P_1, P_2, P_3, P_4, P_5 be the given forces in fig. (i). The method is a general one, but five forces are a sufficient number for consideration.

From any point L in fig. (ii) draw LB_{12} to represent P_1, $B_{12}B_{23}$ to represent P_2 and so on, $B_{45}M$ representing P_5. Join LM to complete the polygon. This diagram (fig. (ii)) is called the Force Diagram and it will be shewn that LM represents the resultant force in magnitude and direction.

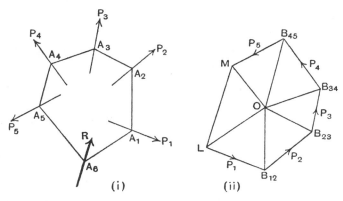

(i)   (ii)

Now take any point O in the plane and join it to the corners of the polygon fig. (ii). O is called the *pole* of the force diagram.

Then return to fig. (i) and take any point A_1 on the line of action of the force P_1 and construct a polygon thus:

Draw

A_1A_2 parallel to $B_{12}O$ to meet the line of action of P_2 in A_2,
A_2A_3 parallel to $B_{23}O$ to meet the line of action of P_3 in A_3,
A_3A_4 parallel to $B_{34}O$ to meet the line of action of P_4 in A_4,
A_4A_5 parallel to $B_{45}O$ to meet the line of action of P_5 in A_5;

lastly, from A_5 draw a parallel to MO, and from A_1 a parallel to OL to meet in A_6. Then we shall prove that the resultant force is equal and parallel to LM and acts through A_6.

The proof is as follows:

The force P_1 acting at $A_1 = \overline{LB_{12}}$
$$= \overline{LO} \text{ along } A_6A_1 + \overline{OB_{12}} \text{ along } A_2A_1;$$
The force P_2 acting at $A_2 = \overline{B_{12}B_{23}}$
$$= \overline{B_{12}O} \text{ along } A_1A_2 + \overline{OB_{23}} \text{ along } A_3A_2;$$
The force P_3 acting at $A_3 = \overline{B_{23}B_{34}}$
$$= \overline{B_{23}O} \text{ along } A_2A_3 + \overline{OB_{34}} \text{ along } A_4A_3;$$
The force P_4 acting at $A_4 = \overline{B_{34}B_{45}}$
$$= \overline{B_{34}O} \text{ along } A_3A_4 + \overline{OB_{45}} \text{ along } A_5A_4;$$
The force P_5 acting at $A_5 = \overline{B_{45}M}$
$$= \overline{B_{45}O} \text{ along } A_4A_5 + \overline{OM} \text{ along } A_6A_5.$$

With the exception of the first and last, these forces are all equal and opposite in pairs round the sides of the polygon A_1, A_2, A_3, ... There remain only the forces $\overline{LO}$ along A_6A_1 and $\overline{OM}$ along A_6A_5, and these have a resultant R represented in magnitude and direction by $\overline{LM}$ and acting through the point A_6.

The polygon A_1, A_2, ... A_6 is called a **Funicular Polygon.** It is assumed that the given forces are acting upon the same rigid body, but it is clear that if the points A_1, A_2, ... A_6 were joined by threads or by smoothly jointed light rods forming the sides of a polygon the force R reversed in direction and the given forces P_1, P_2, ... would maintain the polygon in equilibrium.

8·12. *Special case of a system equivalent to a couple.* When the force polygon is closed by the coincidence of the points L, M, the resultant force is zero, but we are not entitled to assume that the given forces are in equilibrium. We must carefully construct a funicular polygon; then, since L and M are coincident, the line through A_5 parallel to MO and the line through A_1 parallel to OL are either coincident or parallel. In the former case the force system reduces to two equal and opposite forces LO, OM in the same straight line and the system is in equilibrium, and in the latter case the system reduces to a couple of equal and opposite forces LO, OM in parallel lines, and the moment of the couple is $LO \times$ distance between these parallel lines.

8·2. It is evident that for a given system of forces by taking different positions for the pole of the force diagram a great variety of funicular polygons can be constructed and we have the following theorem:

If the pole of a force diagram moves along a straight line the sides of the funicular polygon turn about fixed points which all lie on a parallel straight line.

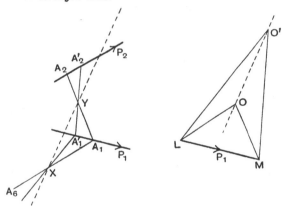

Let LM be the side of the force diagram that represents the force P_1, and when O is the pole let A_6A_1, A_1A_2 be sides of the funicular polygon that meet on the force P_1, parallel respectively to LO, OM.

Let O' be another position of the pole and through any point X on A_6A_1 draw XA_1' parallel to LO' and through A_1' draw $A_1'A_2'$ parallel to $O'M$ meeting A_1A_2 in Y. Then the force P_1 is equal either to

$\overline{LO}$ acting along XA_1 and $\overline{OM}$ acting along YA_1,

or to

$\overline{LO'}$ acting along XA_1' and $\overline{O'M}$ acting along YA_1'.

Therefore these two pairs of forces are equivalent, from which we deduce that

$$\overline{O'L} + \overline{LO} \text{ acting at } X \equiv \overline{O'M} + \overline{MO} \text{ acting at } Y,$$

or $\qquad \overline{O'O}$ acting at $X \equiv \overline{O'O}$ acting at Y,

and this can only be true if XY is parallel to OO'.

Similarly, if the sides A_2A_3 and $A_2'A_3'$ of the two funicular polygons which have O and O' as poles intersect in Z, we can prove that YZ is parallel to OO', and therefore X, Y, Z, ... are collinear points and the theorem follows. It is clear moreover that the choice of the point X on the line A_6A_1 is arbitrary, but having chosen a point X then as O moves along a given line OO', the sides of the funicular polygon will all turn about their intersections with a parallel to OO' drawn through X.

8·21. Reverting to the figures of **8·1**, for different positions of the pole O, the lines A_1A_6 and A_5A_6 always intersect in a point on the line of action of the resultant force; so that for different positions of the pole the locus of the intersection of a pair of sides of the funicular polygon is a straight line. This is true of any pair of sides, for example A_1A_2 and A_4A_5 intersect on the line of action of the resultant of the forces P_2, P_3, P_4.

8·22. *For a given system of forces to construct a funicular polygon two of whose sides shall pass through given points.*

Suppose that the sides A_1A_2 and A_4A_5 are to pass through given points X, Y. We have seen (**8·21**) that these two sides intersect on a fixed line, viz. the resultant of P_2, P_3, P_4. Let SS' be this line. It is parallel to $B_{12}B_{45}$ in the force diagram. On it take any point K and join KX, KY.

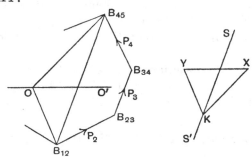

Then, in the force diagram, draw parallels to KY and XK from B_{12}

and B_{45} intersecting in O, and O is the pole for a funicular polygon satisfying the required conditions. For the resultant of the forces P_2, P_3, P_4 is $\overline{B_{12}B_{45}}$ acting along $S'S$ and this is equivalent to forces $B_{12}O$ acting along KY and $\overline{OB_{45}}$ acting along KX.

For different points K on SS' the locus of O is a straight line OO' parallel to YX (8·2).

8·3. Parallel Forces. Exactly the same method applies when the given forces are parallel, with the simplification that the sides of the force diagram all lie in the same straight line.

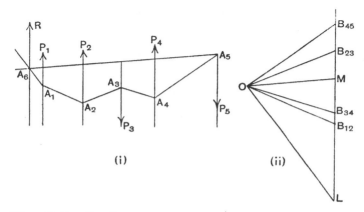

(i)   (ii)

Thus if P_1, P_2, P_3, P_4, P_5 are five parallel forces acting in given lines, we draw LB_{12}, $B_{12}B_{23}$, $B_{23}B_{34}$, $B_{34}B_{45}$, $B_{45}M$ to represent them in order, take any point O for the pole of the force diagram and join it to the points L, B_{12}, B_{23}, ... M. Taking any point A_1 on the line of action of P_1 we construct a funicular polygon $A_1A_2, ... A_6$ as in **8·1**, and it follows by the same proof as in **8·1** that the resultant of the five given forces is a force R of magnitude and direction LM acting through A_6.

8·31. *Given a system of parallel forces to determine the magnitudes of two parallel forces which shall act in given straight lines and be together equivalent to the given system.*

Let P_1, P_2, P_3 be the given forces and XX', YY' the given straight lines. It is sufficient to consider a system of three forces as the same method will apply for any number. Construct a force diagram LB_{12}, $B_{12}B_{23}$, $B_{23}M$ to represent the forces and take any point O as pole. From any point K on XX' draw KA_1 parallel to OL, then A_1A_2 parallel to OB_{12}, A_2A_3 parallel to OB_{23} and A_3H parallel to OM to meet YY' in H as in fig. (i). Join KH and from O in fig. (ii) draw ON parallel to KH to meet LM in N. Then we have the

Force P_1 at $A_1 = \overline{LB_{12}}$
$$= \overline{LO} \text{ along } A_1K + \overline{OB_{12}} \text{ along } A_1A_2;$$
Force P_2 at $A_2 = \overline{B_{12}B_{23}}$
$$= \overline{B_{12}O} \text{ along } A_2A_1 + \overline{OB_{23}} \text{ along } A_2A_3;$$
Force P_3 at $A_3 = \overline{B_{23}M}$
$$= \overline{B_{23}O} \text{ along } A_3A_2 + \overline{OM} \text{ along } A_3H.$$

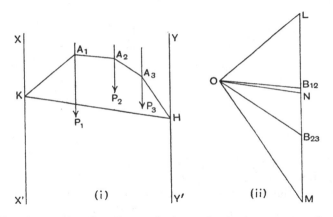

The forces in the lines A_1A_2 and A_2A_3 are equal and opposite pairs, and there remain the forces

$$\overline{LO} \text{ along } A_1K \quad \text{and} \quad \overline{OM} \text{ along } A_3H.$$

We may suppose these to act at K and H respectively, and write $\overline{LO}$ acting at $K = \overline{LN}$ along $XX' + \overline{NO}$ along HK

and $\overline{OM}$ acting at $H = \overline{ON}$ along $KH + \overline{NM}$ along YY'.

The forces in the line KH are equal and opposite, so that the system is equivalent to a force LN along XX' and a force NM along YY'.

The method established above is important for it enables us

to find graphically the supporting forces when a beam is supported at two points and carries given loads at other points.

8·32. Example. *Given a system of parallel forces* P_1, P_2, P_3, P_4, P_5, *to construct a funicular polygon to pass through two given points, X on the line of action of* P_1 *and Y on the line of action of* P_4.

Draw a force diagram LM for the five forces with any point O as pole, and starting from X on P_1 draw consecutive sides of a funicular polygon $XA_2A_3A_4A_5$. Then a line OK parallel to XA_4 meets LM in K and the

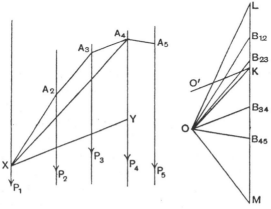

four forces P_1, P_2, P_3, P_4 are equivalent to a force LK at X and KB_{45} at A_4. Draw KO' parallel to YX, then if we construct a funicular using any point O' on KO' as pole starting from X it will pass through Y, for since $O'K$ is parallel to XY we shall have the four forces P_1, P_2, P_3, P_4 equivalent to a force LK at X and KB_{45} at Y.

8·4. *Graphical Representation of Bending Moment.* A measure of the moment of a force or system of forces about a point may be obtained from the funicular polygon.

Let P be a force represented in a force diagram by the line ML. Let O be the pole of the force diagram and AB, AC the sides of the funicular polygon parallel to OL, OM. Then if we require the moment of P about a point X,

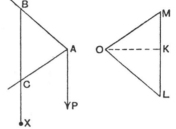

we draw through X a parallel to LM cutting the sides of the funicular polygon in B and C.

Now the force P or $\overline{ML}$ is equivalent to $\overline{MO}$ along AC and $\overline{OL}$ along BA. Therefore the moment of P about $X =$ sum of moments about X of $\overline{MO}$ along AC and $\overline{OL}$ along BA. Now if OK be perpendicular to ML and we regard the last two named forces as acting at C and B respectively, in calculating their moments about X we need only consider their components at right angles to BC which are $\overline{KO}$ and $\overline{OK}$ respectively. Therefore the moment of P about X

$$= \overline{KO}.CX + \overline{OK}.BX$$
$$= \overline{OK}.BC.$$

And $\overline{OK}$ is the same for all positions of X, so that the required moment is proportional to the intercept BC.

In like manner the sum of the moments about X of any system of forces may be measured. For the sum of their moments = the moment about X of their resultant; and the first and last sides of the funicular polygon intersect on the resultant, so if we take P to represent the resultant and draw a parallel to it through X cutting the first and last sides of the funicular polygon in B and C the sum of the moments of the forces about X will be proportional to the intercept BC.

8·41. Examples. (i) *As an application of the method of* **8·4** *consider the case of a horizontal beam supported at its ends and carrying loads* P_1, P_2, P_3 *at given points.*

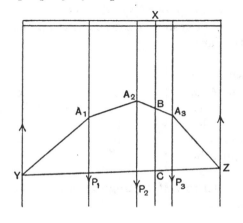
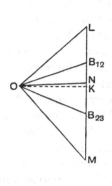

We construct a force diagram $LB_{12}B_{23}M$ with a pole O and a funicular polygon $YA_1A_2A_3Z$ as in **8·31**; the line ON parallel to YZ determining the supporting forces NL and MN which act through Y and Z. To find the bending moment at any point X on the beam, let the vertical through X cut the sides of the funicular polygon in B, C; then the bending moment at X is proportional to BC and is measured by $BC \times OK$, where OK is the perpendicular from the pole to the line LM in the force diagram. For the bending moment about X is the sum of the moments about X of the forces at the corners A_3 and Z of the funicular polygon, and, in finding graphically the resultant of these two forces, the first and last sides of the corresponding funicular polygon would be A_2A_3 and ZY.

(ii) *Two beams ABC, CD are smoothly jointed at C and supported horizontally at A, B and D. Determine graphically the supporting forces at A, B and D for given loads P_1 and P_2, and represent the bending moment graphically.*

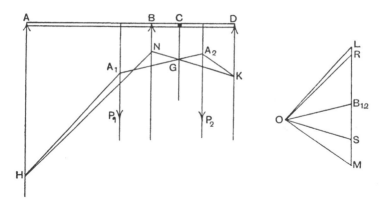

Draw the force diagram $LB_{12}M$ so that LB_{12} and $B_{12}M$ represent the forces P_1 and P_2. Take the verticals through A, B, C, D and construct a funicular polygon by the following steps: from any point H on the vertical through A draw HA_1 parallel to OL to meet P_1 in A_1, then A_1A_2 parallel to OB_{12} to meet P_2 in A_2, and A_2K parallel to OM to meet the vertical through D in K. We then make use of the fact that there is a smooth joint and therefore no bending moment at C, so that the funicular polygon must cross itself on the vertical through C. Let A_1A_2 cut this vertical in G. Join KG cutting the vertical through B in N, and complete the funicular polygon by joining NH. In the force diagram draw OR, OS parallel to HN, NK.

Then the reactions at A, B and D are equal to RL, SR and MS respectively. This may be proved as follows:

$$P_1 + P_2 = \overline{LB}_{12} + \overline{B_{12}M}$$
$$= \overline{LO} \text{ along } A_1H + \overline{OB}_{12} \text{ along } A_1A_2$$
$$+ \overline{B_{12}O} \text{ along } A_2A_1 + \overline{OM} \text{ along } A_2K$$
$$= \overline{LO} \text{ acting at } H + \overline{OM} \text{ acting at } K$$
$$= \overline{LR} \text{ along } AH + \overline{RO} \text{ along } NH$$
$$+ \overline{OS} \text{ along } NK + \overline{SM} \text{ along } DK$$
$$= \overline{LR} \text{ along } AH + \overline{RS} \text{ along } BN + \overline{SM} \text{ along } DK,$$

and the reactions are equal and opposite to these three forces.

The bending moment at any point of the beam will be proportional to the intercept made by the funicular polygon on the vertical line through the point.

We notice that in addition to vanishing and changing sign as we pass the joint C, the bending moment also vanishes and changes sign at a point between B and the force P_1.

8·5. Reciprocal Figures.
A force diagram and the corresponding figure containing the forces and funicular polygon are so related that to every line in one figure there corresponds a parallel line in the other figure. This is a reciprocal property, and is one of the properties of a class of figures to which the name *Reciprocal Figures* was given by Clerk Maxwell[*] who defined them thus:

"Two plane figures are reciprocal when they consist of an equal number of lines, so that corresponding lines in the two figures are parallel, and corresponding lines which converge to a point in one figure form a closed polygon in the other.

"Since every polygon in one figure has three or more sides, every point in the other figure must have three or more lines converging to it; and since every line in the one figure has two and only two extremities to which lines converge, every line in the other figure must belong to two and only two closed polygons."

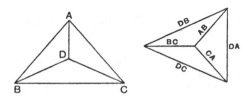

The simplest plane figure fulfilling the conditions is that formed by the six lines which join four points in pairs as shewn in the two figures above. The reciprocal figure consists of six lines parallel to the original six, points in one figure corresponding to triangles in the other.

* *Scientific Papers*, vol. I, p. 514.

8·51. Not every figure has a reciprocal. It is necessary that the same relation shall hold between the numbers of points, polygons and sides as between the numbers of corners, faces and edges of a polyhedron, viz. $C + F = E + 2$, where C, F and E denote the numbers of corners, faces and edges: in fact that the figure may be regarded as the plane projection of a polyhedron.

In the above figures there are four corners, four triangles and six sides, so that $C + F = E + 2$, and this is the simplest pair of reciprocal figures. If we derive another such figure by drawing additional lines, the effect of drawing a line from a corner H to a point K unconnected with the figure is to increase both C and E by unity; but if we draw a line from one corner H to another corner K, while adding nothing to C, we increase both F and E by unity. It follows that the relation $C + F = E + 2$ continues to hold for all figures built up in this way.

A simple way of constructing a reciprocal to a figure composed of triangles is to join the centres of the circumscribing circles of all the triangles and then turn the figure through a right angle.

8·6. Frameworks. By a framework we mean a number of bars or bodies of any shape jointed together. We shall assume that the framework lies in one plane and that the joints are formed by smooth pins passing through small holes in the bodies, a 'single' joint being a point where two bodies only are connected, at a 'double' joint three bodies are connected and so on. Each bar or body is connected with the others at two points only. The external forces are applied to the pins so that the only forces acting on a body are the two forces at its joints and these must be equal and opposite, so the body may be adequately represented as a bar connecting the two joints.

8·61. A framework is said to be *stiff* or *rigid* if it cannot be deformed. For example, a triangle of jointed rods is a stiff frame; a quadrilateral of four rods AB, BC, CD, DA is not

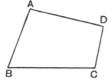 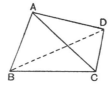

stiff, but can be made so by the addition of a diagonal rod AC; while two diagonals AC and BD would make it *overstiff*, for if the length of BD, say, were such as to cause a tension or thrust between the points B and D there would be proportional forces along all the other rods (**6·43**(i)). A framework may be described as *just stiff* when the removal of a single bar would render it deformable. There is a connection between the number of joints of a framework and the number of bars necessary to make the framework stiff. Let us begin with a single bar connecting two points A and B; two more bars are required to connect a third point rigidly with A and B, and so on for every joint added to the framework two bars must be added. Hence if the framework contains n joints in all, i.e. $n - 2$ in addition to A and B, the number of bars necessary to make it stiff is $1 + 2(n - 2) = 2n - 3$. Otherwise, using plane co-ordinates, to fix the position of n points in a plane, $2n$ co-ordinates must be given; but this would fix not only the shape of the framework but also its position in the plane, and since only its shape is to be fixed and it can move as a whole with three degrees of freedom, therefore $2n - 3$ data are required to determine its shape, and these would be provided by the lengths of $2n - 3$ bars, so that $2n - 3$ bars are sufficient to make the framework stiff.

8·62. Determination of Stresses in a Framework. Consider a framework of n joints and $2n - 3$ bars so that the framework is just stiff. By resolving in two directions for the forces at each joint we obtain $2n$ equations connecting the stresses in the bars and the given external forces; but the external forces alone have to satisfy three conditions of equilibrium, which must be embodied in the $2n$ equations already obtained. Therefore there are $2n - 3$ independent equations for the determination of the stresses in the $2n - 3$ bars, and the problem of finding the stresses in a framework which is just stiff is in general a determinate one. There may, however, be exceptional cases in which the $2n - 3$ equations are not independent or are inconsistent and the problem is then indeterminate.

For example in the case of three bars AB, BC, CA smoothly jointed and such that

$$AB = AC + CB.$$

The jointed structure would be in equilibrium under the action of equal and opposite forces P applied to the pins at A and B. But if the stresses in AC, AB and BC are denoted by R, S and T, we have

$$R + S = P, \quad R = T \quad \text{and} \quad S + T = P$$

and learn no more by resolving in any other direction, so that the stresses are in this case indeterminate.

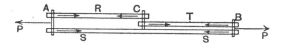

If a framework is over-rigid the stresses are indeterminate because the number of unknown stresses exceeds the number of independent equations, and, as in the example of the over-stiff quadrilateral (**6·43** (i)), the members may be in a state of stress without the application of external forces. Such a framework is said to be *self-stressed*. The three rods in the last figure might constitute a self-stressed frame if, say, AB were in a state of thrust S, while the other two rods were in a state of tension. The equations at the joints being $S = R$, $R = T$ and $T = S$, and the actual values indeterminate.

8·63. The Problem of a Framework. A framework is in equilibrium under the action of given external forces and possibly also certain constraints; for example, it may carry certain loads attached to some of its joints and it may be supported or constrained at other joints; but, as explained in **6·2**, the constraints must not involve more than three unknown elements or the problem is indeterminate. It is required to determine the stresses in the different members of the framework. We shall assume in the first place that the problem is determinate, in that it is possible to find by graphical or analytical methods all the supporting forces or forces of constraint. We shall also assume that the framework is not over-stiff and that at least one of the joints at which an external force is applied is a single joint. The stresses in the members meeting in this joint can be determined by a triangle of forces; we then proceed to the joints at the other ends of these members and can construct

triangles or polygons of forces for these joints provided the
members that meet in a joint do not involve more than two
unknown stresses, and continue the process until all the stresses
are found. In practice it is not necessary to construct separate
triangles or polygons for the different joints after the first, but
merely to add, for each joint, additional lines to the figure
already obtained. The figure constructed will in general be
reciprocal to the figure consisting of the framework and the
external and constraining forces, in that it will contain the
same number of lines such that corresponding lines are parallel,
and corresponding to lines which meet in a point in the framework there will be lines forming a closed polygon in the other
figure. There is not, however, complete reciprocity in the sense
defined by Maxwell because the lines that represent external
forces applied to the framework are not sides of closed polygons.

The method will be illustrated by a few examples. It will be
found that there is sufficient reciprocity between the two
figures for us to adopt a method of lettering introduced by
R. H. Bow in 1873. The lines that represent external forces or
supporting forces are placed outside the framework and, if we
imagine them to be produced outwards to the edges of the paper,
we see the latter divided up into triangles or polygons by the
members of the framework and the forces applied to it. We
then designate each polygonal area by a letter. If a line separates two areas A and B, then we place the letters A and B at
the ends of the corresponding line in the force diagram, and so
we obtain a complete correspondence between the lettering of
the diagrams, showing the way in which a polygonal area in
one figure corresponds to a point in the others.

8·64. Examples. (i) *Fig.* (i) *represents a framework, hinged at
O, further supported by a horizontal force P of unknown magnitude at
the corner above O and carrying a single load W. To find P, the force
at the hinge O and the stresses in the members of the framework.*

Firstly we letter the framework by the method described in **8·63**, so
that each line in fig. (i) has a lettered region on each side of it. The joint
to which W is attached is a single joint and we construct first a triangle
of forces FED for this joint taking any convenient length FE to
represent W. It is important to notice at this stage that this triangle of
forces is named FED (*not FDE*), because FE is the sense of the force
W, and that ED, DF give not only the magnitudes but also the senses

in which the forces act along the rods at the joint FED in the framework. We mark these senses by inserting arrow heads in the framework, upwards in ED and right to left in DF. We then insert arrow heads to indicate the equal and opposite forces which act at the other ends of the same rods.

Then moving along the framework we notice that at the upper joint $FDCB$ there are four rods and only one known force, viz. that in FD, so we are not yet in a position to deal with this joint. But at the joint below, viz. DEC, there are only two unknown forces, viz. those in EC and CD; so we construct a triangle of forces for this point, starting with the known force DE, then drawing EC horizontal and CD vertical; and then put arrow heads to indicate the senses in which the forces act at this joint, viz. EC left to right and CD upwards, and in the contrary directions at the opposite ends of the rods.

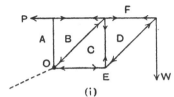 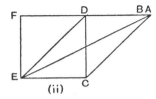

(i)   (ii)

We may now proceed to the joint $FDCB$, where there are now only two unknown forces CB and BF; we construct a polygon of forces for this joint beginning with the known forces FD, DC, then drawing CB and BF in the required directions and noting that two sides FD, BF of the polygon overlap. We then put in arrow heads in the framework to indicate the senses in which the newly determined forces act at this joint, viz. CB upwards and BF right to left, and in the contrary directions at the other ends of the rods.

Proceeding next to the joint above the hinge, we see that there are two horizontal forces AF (or P) and FB and a vertical force BA, and, since the latter would be unbalanced, it must be zero. So in the force diagram the point A coincides with B and AF (or P) is equal and opposite to FB.

There is now only one unknown force, the reaction at the hinge O. Its line of action separates the regions E and A and the polygon of forces for this joint is $ABCEA$, so that the reaction is EA in magnitude, direction and sense.

The arrow heads now serve to shew that the rods BC, DE, EC are in a state of thrust, while BF, FD and DC are in a state of tension.

It would have been difficult to make a blunder in this very easy example, but in more complicated cases difficulties may arise in constructing the force diagram through failure to adopt a consistent convention in the order in which the different members that meet in a joint are taken. Thus in this example a clockwise convention was adopted at every *joint*, viz. FED, DEC, $FDCB$ and so on.

It should be noted that in this and the following examples the arrow heads in the members of the framework are only inserted step by step as their senses are determined by the construction of the force diagrams.

(ii) *The rough sketch represents a jointed framework of light bars, loaded as indicated; AC, CE are horizontal and the angles at A, C, E are each 45°; the lengths AC, CE are each 20 ft. and CF is 16 ft.*

Calculate the reactions at A, E assuming them to be vertical; and draw a force diagram to give the stresses in the bars, distinguishing the ties (in tension) from the struts (under compression). [C.]

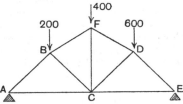

By taking moments about E or A we find that the supporting forces are 500 and 700 at A and E respectively.

We first re-letter the framework according to Bow's method. When a figure like the given one has to be copied, the simplest method is to prick through the corners with a pin on to the paper to be used.

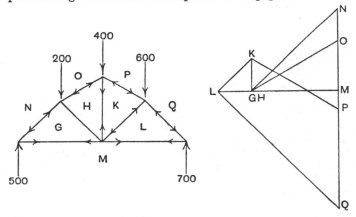

The force diagram is then constructed, taking the joints in the following order: MNG, $GNOH$, $OPKH$, $PQLK$ and QML.

We find that the points G, H coincide, so that there is no stress in the bar GH, and by measurement, or, more accurately, by using the theorem of Pythagoras, that the stresses are as follows:

Ties	Struts
$GM = 500$	$NG = 707$
$LM = 700$	$OH = 583$
$HK = 200$	$PK = 583$
	$KL = 283$
	$LQ = 990$

(iii) *The figure represents a framework of smoothly jointed rods which can turn about a pivot at C. It carries loads of 3 lb. at A and N, and is kept in equilibrium by a force P at B parallel to AC. The rods are all either vertical, horizontal, or inclined at 30° to the vertical. Determine the force P graphically or otherwise. Construct a force diagram and find the stresses in the sides of the triangle LMN, stating whether they are tensions or thrusts.*

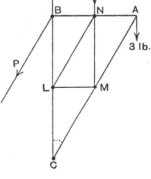

By taking moments about C we find that $P = 3\sqrt{3}$ lb.; or graphically, the total load is 6 lb. bisecting NA, and the resultant of this force and P passes through C, from which it is easy to find the same value for P by a triangle of forces.

After re-lettering the framework we now find that there are two single joints at which we know the external forces, but that the neighbouring joints are all triple joints, so that it is only by making use of all our data that we are able to get a solution. Begin by constructing

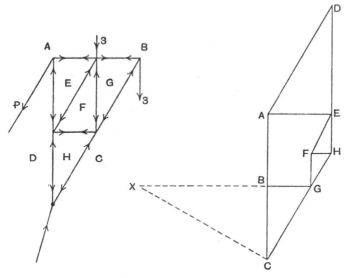

the triangle BCG with the side $BC = 3$ units. Passing next to the centre top joint $ABGFE$, we have found the stress BG and we can draw a line AB in the force diagram to represent the load of 3 lb., but we are unable to complete the pentagon because three sides are of unknown lengths. But if we pass on to the left-hand top joint we know that the line DA is to be $3\sqrt{3}$ units of length and inclined at 30° to the vertical. To construct this length make CBX a right angle and $BCX = 60°$, then $XB = BC \tan 60° = 3\sqrt{3}$. Then in the force diagram make the triangle of forces for the corner DAE, using the point A already obtained and making DA equal to XB. This determines the point E, and we can now return to the joint $ABGFE$ and complete the pentagon, of which we have already got the points G, B, A, E, by drawing the sides GF, FE.

Take next the joint $FGCH$, two of the sides FG, GC of the quadrilateral having already been found; then the joint $DEFH$, for which DE, EF, FH have already been found. The force diagram is now complete, and CD represents the force at the pivot.

The required stresses are found thus from the force diagram

$$CH = AD = 3\sqrt{3}, \text{ and } CG = BC \sec 30° = 2\sqrt{3},$$

therefore $\qquad FE = GH = \sqrt{3}$ and this is a thrust;

then $\qquad FH = \frac{1}{2}FE = \frac{1}{2}\sqrt{3}$ and this is a tension;

lastly $\qquad FG = EH = \dfrac{\sqrt{3}}{2} FE = 1\cdot5$ also a thrust.

8·7. The Method of Sections. If a straight line section be drawn across a framework intersecting a number of the bars, then the stresses in these bars must together be in equilibrium with the external forces or constraints applied to either part into which the framework is divided by this section; and if the cutting line intersects not more than three bars, the stresses in these bars can be found from the equations of equilibrium. We will illustrate the method by a simple example.

The framework shewn in the figure, in which the angles are all either 30° or 120°, is hinged to a wall at A and F and carries a load of 50 lb. at D. To find the stresses in the bars using the method of sections.

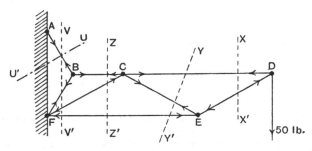

The section XX' which cuts the bars CD, DE gives no more information than that the three forces at D are in equilibrium, and by Lami's Theorem we find that

$$(ED) = 100\,\text{lb.} \quad \text{and} \quad (DC) = 50\sqrt{3}\,\text{lb.},$$

where (ED) denotes the force along ED.

The section YY' gives that the forces in CD, CE and EF balance the load of 50 lb. By resolving vertically we get

$$(EC)\sin 30° = 50, \quad \text{therefore} \quad (EC) = 100\,\text{lb.};$$

and by resolving horizontally, we get

$$(FE) = (EC)\cos 30° + (DC) = 100\sqrt{3}\,\text{lb.}$$

The section ZZ' gives that the forces in BC, CF and FE balance the load of 50 lb. By resolving vertically we get

$$(FC)\sin 30° = 50, \quad \text{therefore} \quad (FC) = 100\,\text{lb.};$$

and by resolving horizontally we get

$$(CB) = (FC)\cos 30° + (FE) = 150\sqrt{3}\,\text{lb.}$$

A section UU', which cuts AB alone, merely serves to point out that the moment of (BA) about F is equal to the moment about F of the load at D, from which we find that

$$(BA) = 150\sqrt{3}\,\text{lb.}$$

Finally a section VV', which cuts AB, BF, CF and EF, gives that the four forces in these bars balance the load of 50 lb. at D; so that by resolving horizontally we get

$$(BF)\cos 60° + (BA)\cos 60° = (FC)\cos 30° + (FE),$$

giving $\qquad\qquad (BF) = 150\sqrt{3}\,\text{lb.}$

We have determined all the stresses by using sections, but the last two forces might have been deduced at once because the angles at B are all 120° so that
$$(BF) = (BA) = (BC).$$

We have said nothing about the senses in which the stresses act. These are easily determined from the fact that they have to be such as to keep the joints in equilibrium; and it is well to bear in mind that in every case we were considering the equilibrium of the portion of the frame on the right of the section, so that the reactions were always from or towards joints on the right of the section.

8·8. Distributed Loads. In the foregoing problems we have only considered cases in which the loads are applied at the joints. When it is required to take into account the weights of the bars or of loads distributed along the bars we may proceed as follows: resolve the total load on each bar into two forces acting at the joints at the ends of the bar and deduce by the foregoing methods a system of stresses in the bars which together with these concentrated loads would maintain all the joints in equilibrium. Then to find the tension or thrust, shearing force and bending moment in any bar AB under the given distributed loads we argue thus:

We want to find the reactions of the rest of the system on the bar AB and these reactions are independent of whether the load on AB is a distributed load or represented by its concentrated components, say P at A and Q at B. Suppose that the stress in AB, in the problem of concentrated loads, is a thrust T and that the resultant reaction of the rest of the system on AB at A is a force R. Then since the joint A is in equilibrium under the action of R, T and P, we have a vector equation

$$\mathbf{R} + \mathbf{P} + \mathbf{T} = 0$$

or $$\mathbf{R} = -\mathbf{P} - \mathbf{T}.$$

Therefore the resultant reaction at A of the rest of the system upon the bar AB is simply the resultant of P and T reversed in direction, similarly the resultant reaction at B is the resultant of Q and T reversed in direction.

8·81. Example. *The pin-jointed framework shewn in the figure is supported by vertical reactions at D and E, and the loads consist of a weight of 2 tons suspended at F and the weight of the curved member BC,*

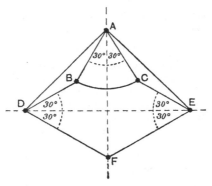

which may be taken as 2 tons, uniformly distributed along the horizontal line BC. Prove that the bending moment at the middle of the member BC is

$$a\,(\sqrt{3} - \tfrac{5}{4})\,\text{tons-feet},$$

where a is the length, in feet, of the members AB, BD, AC, CE, and the centre line of BC is a circular arc of radius a with A as centre.   [T.]

By symmetry it follows that the vertical reactions at D and E are each one-half of the total load, i.e. 2 tons. Also since FD, FE make angles of 60° with the vertical, the stresses in FD and FE are tensions of 2 tons. Then considering the joint at D, where stresses in DB and DA are unknown, and either by constructing a polygon of forces or by resolving, we find that there is a thrust of 2 tons in DB and no stress in DA.

We then proceed to the joint at B, and suppose that half the weight of the member BC acts at each point B and C, i.e. a force of 1 ton vertically downwards. Having removed the load, the only forces that now act upon BC are forces at B and C and these must be equal and opposite, so that in the problem in which loads are concentrated at the joints the reaction T of BC on the joint at B is a horizontal force.

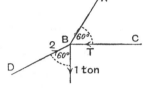

Then, either by constructing a polygon of forces or by resolving vertically for the forces acting at the joint B, it is at once apparent that there is no stress in AB.

If we followed the procedure described in **8·8** we should now reverse the force T and compound with it a force of 1 ton vertically upwards at B, in order to find the reaction at B of the rest of the framework on the member BC when its load is re-distributed. This, however, is unnecessary because in this case the resultant of the two forces last named is clearly the thrust of 2 tons in the member DB.

The member BC is therefore in equilibrium under the action of its weight and the thrusts of 2 tons each in the members DB and EC.

For the bending moment at M, the middle point of the arc BC, we take the moments about M of the forces acting on the portion BM, viz.

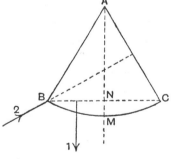

$2 \cos 30° . NM$
$\qquad + 2 \sin 30° . BN - 1.\tfrac{1}{2}BN$
$= a (1 - \cos 30°) \sqrt{3} + \tfrac{1}{2}a - \tfrac{1}{4}a$
$= a (\sqrt{3} - \tfrac{5}{4})$ tons-feet.

This example serves to indicate that, if a member is regarded as perfectly rigid, its curvature makes no difference to the distribution of stress in the other members of the framework but only in itself.

EXAMPLES

1. Four forces acting in one plane are in equilibrium, their lines of action are all given and also the magnitude of one of them; give a geometrical construction to find the magnitudes of the others.   [S.]

2. A light rod AB, 10 ft. long, is supported at A and at another point. A load of 1 lb. is suspended from B, and loads of 5 lb. and 2 lb. at points 2 ft. and 6 ft. from A. (i) If the second support is at the middle point of AB, find graphically the pressures on the supports. (ii) If the pressure on the support at A is required to be 4 lb., find graphically where the second point of support must be.   [S.]

3. A uniform plank of weight 40 lb. and length 8 ft. carries a weight of 12 lb. at a point 3 ft. from one end. If the plank is supported at its ends in a horizontal position, determine *graphically* the pressures on the supports.   [I.]

4. $ABCDE$ is a light rod supported horizontally at B and D and carrying loads of 10, 15 and 20 lb. at A, C and E, where $AB = DE = 2$ ft., $BC = 3$ ft. and $CD = 4$ ft. Determine graphically the pressures on the supports and construct the diagram of bending moments.

5. AC and CF are light rods smoothly jointed at C, supported at B, D and F and carrying loads of 8 and 30 lb. at A and E, where

$$AB = BC = CD = DE = EF.$$

Determine graphically the pressures on the supports and construct the diagram of bending moments.

6. AB, BC, CD are equal light rods smoothly jointed at B and C and supported horizontally at A, G, H, K, where G, H are points of trisection of BC and K is the middle point of CD. They carry loads of 4, 20 and 2 lb. at D and the middle points of BC and AB. Determine graphically the pressures on the supports and construct the diagram of bending moments.

7. A portion of a Warren girder consists of three equilateral triangles ABC, ADC, BCE, the lines AB, DCE being horizontal, and AB uppermost. It rests on vertical supports at D and E, and carries 3 tons at A and 1 ton at B. Find the reactions at the supports and the stresses in the four inclined bars, stating whether they are in tension or compression.   [S.]

8. Five weightless rods AB, BC, CD, DA and BD, smoothly jointed at their ends, form a framework: AB is vertical with B above A, $BAD = 30°$, $ABD = 60°$, BC is horizontal and $BD = DC$. A weight W

is suspended from C and the framework is supported by a horizontal force at B and a force at A. Find the stress in each of the rods, distinguishing between thrust and tension. [S.]

9. A framework is made of light rigid rods smoothly jointed. Its configuration may be obtained by describing squares $ABDE$, $ACFG$ on the sides AB, AC of an equilateral triangle ABC, the squares being external to the triangle, and joining EG, so that EAG is an isosceles triangle with an angle of 120° at A. The points FD are pressed together by external agency with a force of magnitude P: determine, by graphical methods or otherwise, the stress in the rod EG. [S.]

10. $ABCD$ is a rhombus of freely jointed rods in a vertical plane and B, D are connected by a rod jointed to the rhombus. A and B are fixed so that AB is horizontal and below the level of CD. The acute angle A of the rhombus is α. If a weight W is hung from C, draw the force diagram and find the stress in the rod BD in terms of W, α. [S.]

11. A regular pentagon $ABCDE$, formed of light rods, jointed at the angles, is stiffened by two light jointed bars AC, AD. Two equal and opposite forces, each equal to 3 lb. weight, are applied at B and E: find graphically or otherwise the stress in each bar of the framework, stating whether it is tensile or compressive. [S.]

12. A frame of five light rods is formed by two congruent triangles ABC, ABD on the same base AB and on opposite sides of it so that angle $ABC =$ angle $ABD = 45°$ and angle $BAC = 120°$. If this frame be acted upon at C and D by equal and opposite forces, each of magnitude 56 lb. weight, determine by a force diagram the stresses in the rods. [S.]

13. A regular hexagon $ABCDEF$ formed of light rods, connected by smooth joints, stiffened by light rods FB, FC, FD, and suspended from the point A, has weights each equal to W attached to the joints B, C, D, E, F. Find graphically the stress in each rod. [S.]

14. AB, BC, CD are three sides of a square, E is the point of intersection of AC and BD, and $AEDF$ is a square; nine straight rods without weight occupy the positions AB, BC, CD, DF, FA, AE, EC, BE, ED, and are jointed together at their ends; the frame so formed is in a vertical plane the point B being vertically above C, a given weight is attached to F, C is hinged to a fixed point, and equilibrium is maintained by a horizontal force acting at B; find the tension or thrust in each of the rods. [S.]

15. A framework, lying upon a smooth horizontal table, is composed of ten rods. Of these AB, BC, CD, DA form a rhombus with angles of 60° at B, D, and AE, EC, CF, FA form a square of which E is the corner nearest to B. The remaining two rods connect B, E and D, F.

Shew by graphical methods that, if all the joints be smooth and the points A, C be pushed towards one another with force P, the stress in BF and DF will be $\frac{1}{2}P(\sqrt{3}+3)$. Is this a tension or a thrust?   [S.]

16. A rigid plane framework of five jointed bars forming two equilateral triangles BAC, CDA is in equilibrium under the action of three forces on the joints at A, B, D. Prove that the directions of these forces must be parallel or concurrent, lying in the plane of the framework.

If the force at B be 4 cwt. acting perpendicularly to CD and the force at A be in the direction BA, find the force at D and the stresses in the bars by a graphical method.   [S.]

17. ABC is a horizontal line such that $AB=5$ ft. and $BC=15$ ft. D is a point vertically over B such that $BD=10$ ft. E bisects CD. AB, BC, CE, DE, AD, BD, BE are seven freely jointed rods forming a framework. Loads of 10 cwt. each are applied at D and E, and the system is supported at A and C. Find graphically or otherwise the stresses in the various rods, stating which are in tension and which in compression.   [S.]

18. A framework of seven freely jointed light rods is in the form of a regular pentagon $ABCDE$ and its diagonals AC, AD. The framework is in a vertical plane with the lowest rod CD horizontal and is supported at C and D and weights $W, 2W, W$ are suspended at B, A, E respectively. Draw a force diagram to exhibit the stresses in the rods, state which of them are in tension and which in thrust and shew that the tension in CD is
$$2W\tan 18°\,(1-\cos 36°).   \text{[T.]}$$

19. The figure represents a pin-jointed framework of light rods. The rods BC, ED are horizontal and the length of each of the rods AB, BC, ED is equal to AE. The whole is attached to a vertical wall at A and E and the angle XAB is 60°. Determine by graphical methods the stress in each rod.   [S.]

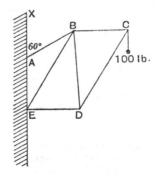

20. Draw a force diagram for the framework shewn in the diagram,

in which all the rods except AB, AC are of the same length. Determine the stresses in AB, AC. [S.]

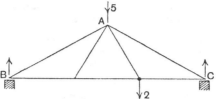

21. The figure shews a roof truss, which is loaded at B with a weight W and is supported at A and C. The angle ABC is a right angle and is trisected by BD and BE; the angles A and C are each 30° and $BA = BC$. Draw a force diagram to shew the tension or compression in each member. [T.]

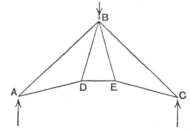

22. The diagram represents a framework of smoothly jointed rods, loaded at CDE, and supported at A and B. If $\alpha = 60°$, draw a diagram giving the stresses in the rods. Shew that if $\tan \alpha = 3$ there is no stress in the rod FD. [S.]

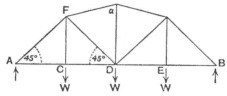

23. The figure represents a stiff framework of light rods freely jointed. The framework is supported at B, C and weights of 3 and 4 lb. are suspended from A, D. Draw a stress diagram for the figure and state whether the stress in each rod is a tension or a thrust. [S.]

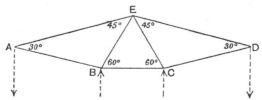

24. The figure represents a Warren girder, which is hinged at A and supported by a smooth horizontal plane at B: equal weights are attached to the points a, b, c. Draw the force diagram, and determine which members are struts and which are ties. [I.]

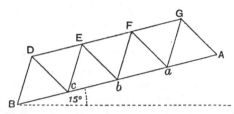

25. Draw a force diagram for the frame sketched, which consists of light rods freely jointed, is supported at A and B, and supports weights of 1, 2 and 5 tons at the corners C, D and E respectively. [T.]

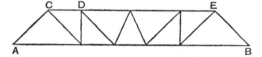

26. In the jointed frame of light rods shewn below, equal and opposite forces are applied at A and B in the line AB. Draw a force diagram for the frame, and state which members are in compression and which in tension. [S.]

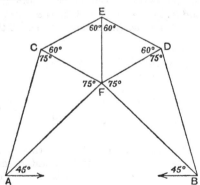

27. A parallelogram $ABCD$ formed of heavy uniform rods freely jointed at their extremities is suspended by the joint A, and kept from collapsing by a weightless rod forming the diagonal BD; prove that the thrust in BD equals $(W + W')\,BD/AC$, where W and W' are the weights of AB and AD respectively. Find also the magnitude of the action at C. [I.]

28. A regular hexagon formed of six uniform freely jointed rods is hung up by one corner and held in shape by horizontal forces R applied at the upper ends of the vertical sides and similar forces S applied at the lower ends. Shew graphically that R is five times S and hence find where to place a light bar horizontally which will alone preserve the shape. [C.]

29. A regular hexagon $ABCDEF$ is formed of heavy rods. A, B, C, E, F are joined to the centre O by light rods and the hexagon is hung from A. Draw the force polygon and determine the thrusts in the light rods. [I.]

30. A symmetrical arch of span a and height h is to be constructed of straight massless jointed rods, to carry seven equal weights w at horizontal distances $\frac{1}{8}a$ apart, in such a way that there shall be no bending moment at any point of any rod. Shew how the form of the arch may be determined by graphical construction, and prove that the horizontal forces necessary to keep the ends in position are aw/h. [C.]

ANSWERS

2. (i) 1·6 lb., 6·4 lb.; (ii) 8 ft. from A. 3. 27·5 lb., 24·5 lb.
4. 15·7 lb., 29·3 lb. 5. 16 lb., 3 lb., 19 lb.
6. 1 lb., 16 lb., 1 lb., 8 lb.
7. 2·5 tons at D, 1·5 tons at E. Thrusts $AD = 2·88$ tons, $AC = ·58$ ton, $BE = 1·73$ tons; tension $BC = ·58$ ton.
8. Tensions $AB = \frac{1}{2}W$, $BC = \sqrt{3}W$; thrusts $AD = \sqrt{3}W$, $CD = 2W$, $BD = W$. 9. P. 10. $W \cos \alpha \sec \frac{1}{2}\alpha$.
11. Tensions $AC = AD = 1·85$ lb.; thrusts $AB = AE = 3$ lb., $CD = 1·15$ lb., $BC = 1·85$ lb.
12. Tensions $AB = AC = AD = 153$ lb.; thrusts $BC = BD = 108$ lb.
13. Tensions $AB = AF = 5W$, $BC = 1·5W$, $CD = ·5W$, $EF = W$; thrusts $BF = 4·33W$, $CF = ·5W$, $DF = ·87W$.
14. Tensions $AF = DE = BE = W/\sqrt{2}$, $AB = W$; thrusts $CD = W$, $BC = \frac{1}{2}W$, $AE = EC = DF = W/\sqrt{2}$. 15. Thrust.
16. 8 cwt. Tensions $AB = \frac{4}{3}\sqrt{3}$ cwt., $AC = \frac{8}{3}\sqrt{3}$ cwt.; thrusts $BC = CD = DA = \frac{2}{3}\sqrt{3}$ cwt.
17. Tensions $AB = 5·625$, $BC = 13·125$, $BD = 5$ cwt.; thrusts $AD = 12·5$, $DE = 6·75$, $EC = 15·77$, $BE = 9$ cwt.
18. Tensions AB, CD, EA; thrusts BC, AC, AD, DE.
19. Tensions $AB = 200$, $BC = 57·7$, $BD = 100$; thrusts $DE = 57·7$, $CD = 115·4$, $BE = 230·8$ lb. 20. Thrusts $AB = 6\frac{1}{3}$, $AC = 7\frac{3}{4}$.
23. Tensions AE, ED; thrusts AB, BC, CD, BE, CE.
24. Tensions Aa, ab, bc, cB, cD, bE, aG, bF; thrusts BD, DE, EF, FG, GA, cE, aF.
26. Tensions AC, CE, ED, DB; thrusts FA, FC, FE, FD, FB.
27. $(W^2 . AD^2 + W'^2 . AB^2 - 2WW'AB . AD \cos BAD)^{\frac{1}{2}}/2AC$.
29. Tension $OA = 2W$; thrusts $OB = OF = 3W$, $OC = OE = W$, where $W =$ weight of a rod.

Chapter IX

FRICTION

9·1. Friction between two bodies in contact at a point was defined in **3·7** as that part of the mutual reaction between the bodies which lies in the tangent plane at the point of contact. When two bodies are in contact over a plane area on the surface of each body, the mutual reaction between them is in general a force acting at some point of this area, and its component, if any, in the plane of contact is called Friction.

The mutual reactions between bodies are *passive forces* in the sense that they only exist because of other forces applied to the bodies. Thus a body may stand alongside a vertical wall and in contact with it without any pressure between them until an external force presses the body against the wall and then there is a mutual reaction between the body and the wall and its amount is no more than is necessary to balance the externally applied force.

Friction, being a component of the mutual reaction, is a passive force. It prevents or tends to prevent the motion of one body across the surface of another, i.e. the relative motion of the points of contact.

9·11. Laws of Friction:

(i) Friction acts in the direction opposite to what would be the direction of relative motion if the friction did not exist; or, in the case of relative motion, the friction opposes the relative motion.

(ii) The magnitude of the friction is always just sufficient to preserve equilibrium or to prevent relative motion of the points of contact, provided that sufficient friction can be obtained.

There is a *limit*, however, to the amount of friction that can be called into play; when this limit is reached the friction is called *limiting friction*, and limiting friction is controlled by the following further laws:

(iii) Limiting friction between two bodies bears a constant ratio to the normal reaction between them; i.e. if R denotes the normal reaction, then μR denotes the limiting friction, where μ is a constant depending on the nature of the materials in contact and called *the coefficient of friction*.

(iv) Limiting friction is independent of the areas in contact.

(v) When motion takes place the foregoing laws of limiting friction are still true, but the coefficient of friction μ is slightly less for bodies in motion than for the same bodies at rest; and the friction is independent of the velocity.

9·12. The foregoing laws are based upon experiments. It must not be supposed that they are rigorously true under all conditions. If the pressure between two bodies were increased to such an extent as to crush the parts of the bodies in contact, the coefficient of friction would not remain invariable.

9·13. Experimental Verifications. A box resting on a horizontal plane is fastened to a string which passes, at first horizontally, over a smooth pulley and carries a scale pan.

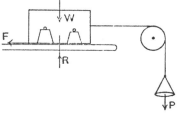

Weights W are placed in the box and a gradually increasing load P of sand or other suitable material can be placed in the scale pan and subsequently weighed.

There is then a normal reaction R equal to W, and a frictional force F equal to P. The load P can be increased until the box just slides, and a verification obtained that in limiting equilibrium F/R is independent of R. By using boxes of different sizes it can be verified that μ is independent of the areas in contact. Results so obtained will not be very precise because it is impossible to eliminate friction entirely from the pulley.

9·14. A second method of experimenting is to place a box containing weights on a horizontal plane which can be tilted about a horizontal line. It will be found that no matter what may be the total weight W the angle λ through which the plane can be tilted before sliding begins is always the same for the same two

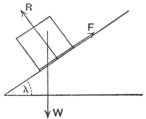

substances in contact. Then if R be the normal reaction and F the friction, by resolving along and perpendicular to the plane, we get

$$F = W \sin \lambda \quad \text{and} \quad R = W \cos \lambda,$$

so that $\qquad\qquad\qquad F/R = \tan \lambda.$

It follows that the coefficient of friction is equal to the tangent of the angle of inclination of the plane to the horizontal when the equilibrium is just about to be broken by sliding.

9·15. Angle of Friction and Cone of Friction. When two

bodies in contact at a point are in limiting equilibrium so that the greatest possible amount of friction is called into play, the angle which the resultant reaction makes with the common normal at the point of contact is called **the angle of friction.**

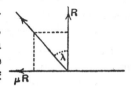

If R denotes the normal reaction, then μR denotes the limiting friction and, if λ is the angle of friction,

$$\tan \lambda = \mu R/R = \mu;$$

so that the angle of friction is the angle whose tangent is the coefficient of friction.

It follows that in **9·14** the greatest angle through which the plane can be tilted before the body slides is the angle of friction as defined above.

The resultant reaction between the bodies is, as explained in **9·1**, a passive force called into play to balance external forces and possibly other constraints. The question arises whether this passive force can adjust itself so as to balance any given force system, and the answer imposes two important limitations:

(i) the resultant of the other forces acting upon either body must pass through the point of contact of the two bodies;

(ii) the same resultant must not make with the common normal at the point of contact an angle greater than the angle of friction, because such a state would imply a component force in the tangent plane whose ratio to the normal pressure exceeded the coefficient of friction, and this is not possible in a state of equilibrium.

It follows that, if, with the common normal NN' to the surfaces as axis, we construct a cone of semivertical angle λ, the angle of friction, then if the resultant of the other forces acting on either body falls within or along the surface of this cone it can be balanced by the reaction of the other body; but, if it falls without the cone, equilibrium is not possible because there is not enough friction available to balance the tangential component of the other forces acting on either body. This cone is called **the cone of friction.**

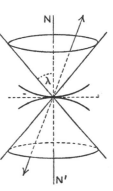

9·16. Numerical Values. The values of the coefficient of friction for various substances have been found by experiment and tabulated:

	About
Timber on stone	0·4
Timber on timber	0·5 to 0·2
Timber on metals	0·6 to 0·2
Metals on metals	0·25 to 0·15
Masonry on dry clay	0·51

These figures are taken from Rankine's *Applied Mechanics*. It will be noticed that all the above values of μ are less than unity. The only substances stated by Rankine to have a coefficient of friction as great as unity are

	earth on earth, damp clay	1·0
and	earth on earth, shingle and gravel	0·81 to 1·11

9·17. Rolling Friction. We saw in 5·1 that a system of coplanar forces acting on a rigid body can always be reduced to a single force acting at a specified point, together with a couple. So far in this chapter we have assumed that, when two bodies are in contact, the other forces acting upon either body apart from their mutual action and reaction are equivalent to a single force acting through the point of contact, and therefore such as could be balanced by the mutual reaction when sufficient friction can be called into play. It is easily conceivable, however, that the system of 'other forces' acting upon either body might not be equivalent to a single force at the point of contact, but to a force and a couple, and in that case the mutual action and reaction between

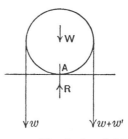

the bodies must consist of a *friction couple* as well as a force in order to maintain equilibrium.

Consider for example a solid uniform cylinder of weight W resting on a horizontal plane, with a cord passing over the cylinder and having weights attached to its ends which may pass through a slit in the plane.

Suppose that the weights suspended from the cord are at first equal to w and that one of them is gradually increased until, when it becomes $w + w'$, equilibrium is just about to be broken by the rolling of the cylinder. The external forces are all vertical and may be represented by a force $W + 2w + w'$ downwards at the point of contact A, together with a couple of moment $w'r$, where r is the radius of the cylinder. The reaction of the plane at A therefore consists of an upward force

$$R = W + 2w + w',$$

and a couple of moment $w'r$, and this couple is called the *friction couple*. If the cylinder were also acted upon by a force with a horizontal component there would also be a friction force along the plane at A.

It is found by experiment that, when equilibrium is about to be broken by rolling, w' varies directly as R and inversely as r, or that the friction couple $w'r$ is proportional to the normal pressure R.

The explanation of the existence of the friction couple is that the cylinder is not perfectly rigid and the contact is really over a small area; that, when rolling is about to begin, at points where the surfaces tend to separate they also tend to adhere, and the normal reactions on the cylinder at these points and at points where the tendency is to produce compression are in opposite senses and so give rise to a couple. This kind of friction is sometimes described as the friction of cohesion.

9·2. Problems. Problems about friction are of very varied kinds. The following example may be taken as typical of a kind of problem in which the coefficient of friction and the configuration are given and it is required to find a certain external force.

A body of weight W is placed on an inclined plane whose inclination α is greater than the angle of friction; to find the least force which will prevent the body from sliding down the plane.

Let R be the resultant reaction of the plane on the body. Since the body tends to slide down the plane the friction on the body acts up the plane, so that in limiting equilibrium R makes an angle λ (the angle of friction) with the normal to the plane on the upper side of the normal. But the normal to the plane makes an angle α with the vertical and $\alpha > \lambda$, therefore R lies be-

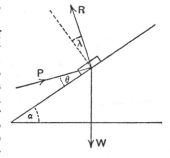

tween the normal and the vertical. Let P be a force which will balance R and W, and let P make an angle θ with the plane.

Then, by Lami's Theorem, we have

$$\frac{P}{\sin(\alpha - \lambda)} = \frac{R}{\cos(\alpha - \theta)} = \frac{W}{\cos(\theta - \lambda)}.$$

It is clear that the values of P and R both depend upon θ, and that P is least when $\cos(\theta - \lambda)$ is greatest, i.e. when $\theta = \lambda$. So the least value of P is $W \sin(\alpha - \lambda)$, and it occurs when P is at right angles to R.

9·21. Another type of problem is one in which the data are a given configuration in limiting equilibrium under given external forces, to find the coefficient of friction.

Example. *A uniform cylinder rests on a horizontal plane with its axis horizontal. A plank lies across the cylinder with one end resting on the plane. Shew that if the plank is just about to slide on the cylinder the angle of friction between them is half the inclination of the plank to the horizontal.*

Let the figure represent a vertical section through the central line of the plank. AB denotes the plank touching the cylinder at D which also touches the horizontal plane at E. There is assumed to be such symmetry that all the forces act in the vertical plane of the diagram.

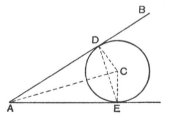

Consider the equilibrium of the cylinder. It is acted upon by three forces: (i) its weight in the vertical line CE, (ii) the reaction of the horizontal plane through E, and (iii) the reaction of the plank at D. Since the first two of these forces pass through E, the third force must also pass through E. Therefore the resultant reaction between the plank and the cylinder acts along DE, but in limiting equilibrium it is inclined to the normal DC at the angle of friction. Therefore CDE is the angle of friction and this is easily seen to be half the angle DAE.

We notice that this result is independent of the weight and length of the plank, so that it would have been less simple to try to obtain a solution by first considering the equilibrium of the plank as this would inevitably have introduced several unknown quantities which would have had to be eliminated.

9·22. Indeterminateness of Problems. Problems concerning the reactions between rough surfaces are indeterminate when more friction is available than is necessary to maintain equilibrium.

Consider for example the case of *a uniform rod with its ends resting on two equally rough inclined planes, in a vertical plane at right angles to the line of intersection of the inclined planes.*

Let AB be the rod. Let the planes make angles α, β with the horizontal and let the normals to the planes at A and B intersect at N. Draw lines at A and B on both sides of the normals making with them angles equal to the angle of friction λ; viz. the lines AML, AHK and BHM, BKL. Then so far as availability of friction goes, the reaction at A may lie anywhere within the angle KAL, and the reaction at B anywhere within the angle MBL: so that these reactions could intersect at any point within the quadrilateral $HKLM$.

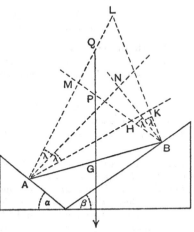

But since there is only one other force acting on the rod, namely its weight acting vertically through its centre of gravity G, therefore the reactions at A and B must intersect on the vertical through G. The only part of this vertical line which falls within the area $HKLM$ is the part PQ in the figure. It follows that for the given position of the rod the reactions at A and B are indeterminate to the extent that they may intersect at any point of the vertical segment PQ, and their ratios to the weight of the rod are then given by Lami's Theorem. It is clear that the position of the rod would not be a possible position of equilibrium if the vertical through G did not meet the area $HKLM$.

In a position of limiting equilibrium the rod must be just about to slide down one plane and up the other, so that the reaction on the rod must be above the normal at one end and below it at the other, i.e. the figure must be such that the vertical through G passes through a corner M or K of the quadrilateral.

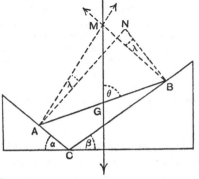

Suppose that the rod is just about to slide down the plane AC and up the plane CB, then the reactions at A and B intersect at M on the vertical through G, and if AN, BN are the normals at A

and B the angles MAN and MBN are both equal to the angle of friction λ.

Then since AN makes an angle α with the vertical, the angle AMG is $\alpha - \lambda$, and for a similar reason BMG is $\beta + \lambda$; and the inclination θ of the rod AB to the vertical is given by

$$\frac{\sin MAG}{\sin AMG} = \frac{MG}{AG} = \frac{MG}{BG} = \frac{\sin MBG}{\sin BMG},$$

or

$$\frac{\sin (\theta - \alpha + \lambda)}{\sin (\alpha - \lambda)} = \frac{\sin (\theta + \beta + \lambda)}{\sin (\beta + \lambda)},$$

which reduces to

$$2 \cot \theta = \sin (2\lambda + \beta - \alpha) \operatorname{cosec} (\alpha - \lambda) \operatorname{cosec} (\beta + \lambda).$$

9·3. Initial Motion. Sometimes a system is in equilibrium under the action of a gradually increasing force and it is required to determine how the system will begin to move when the conditions of equilibrium are no longer satisfied. When the circumstances are such that equilibrium is broken *either* by the turning of one body about a point of contact with another *or* by the sliding of the one over the other, then *the former will happen* if the forces acting on the body reduce to a single force at the point of contact falling within the cone of friction at that point. That this is so follows from the fact that when the body turns the friction required to prevent sliding is less than the maximum friction available, whereas if sliding took place the maximum friction would be called into play; and it is implied in the definition of friction as a passive force that the amount of friction called into play is never more than is necessary to prevent sliding.

We may summarize this conclusion by saying that, when it is a case of rolling or sliding, then rolling takes place, if there is sufficient friction to prevent sliding.

The method of solving such problems is therefore to assume that equilibrium will be broken by the turning of a body about a point and examine whether the reaction at the point lies within or on the cone of friction at the point. If it does so, the assumption is justified. If not, then equilibrium will be broken by sliding.

9·31. Examples. (i) *A heavy cubical block of edge 2a is placed on a rough table with one face parallel to the edge of the table and at a distance $a \cot \alpha$ from it; to the centre of this face a light smooth rod of length l is*

*freely jointed, it passes over the smooth edge of the table and carries a weight
W at its end. Shew that as W is increased the equilibrium of the block is
broken by its tilting about an edge if*

$$\mu > \frac{l\cos\alpha\sin^2\alpha}{a + l\sin\alpha\cos\alpha\,(\sin\alpha - \cos\alpha)}. \qquad [\text{S.}]$$

Let C be the edge of the table and DCB the rod jointed to the block
at B. The forces acting on the
rod are the weight W at D,
the reaction at C at right
angles to the rod since the
edge of the table is smooth,
and the reaction at B. The
two former meet in a point E,
therefore the reaction at B
on the rod is along EB. Let
EB cut the vertical through
the centre of gravity G of the
cube in F. Then since two of
the three forces acting on the

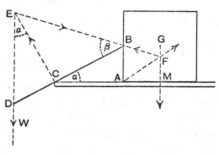

cube, viz. its weight and the reaction at B, pass through F, therefore
the third force, the reaction of the table, must also pass through F. And
from **9·3** equilibrium will be broken by turning about A if the force
along AF falls within the cone of friction at A, i.e. if

$$\mu > \tan BAF.$$

Now $CA = a\cot\alpha$, therefore $BCA = \alpha$; and $DB = l$, therefore
$$DC = l - a\operatorname{cosec}\alpha;$$
and, if $CBE = \beta$,
$$\tan\beta = \frac{EC}{CB} = \frac{(l - a\operatorname{cosec}\alpha)\cot\alpha}{a\operatorname{cosec}\alpha}$$
$$= \frac{l}{a}\cos\alpha - \cot\alpha.$$

Whence we find that
$$\tan(\beta - \alpha) = \frac{l\cos^2\alpha\sin\alpha - a}{l\sin^2\alpha\cos\alpha}.$$

Now
$$\tan BAF = \tan AFM = \frac{a}{FM} = \frac{a}{a - GF}$$
$$= \frac{1}{1 - \tan(\beta - \alpha)}$$
$$= \frac{l\sin^2\alpha\cos\alpha}{a + l\sin\alpha\cos\alpha\,(\sin\alpha - \cos\alpha)},$$

and, if μ is greater than this expression, as W increases the block will
turn about the edge through A. If μ has a smaller value than the fraction
stated, the block will eventually slide.

(ii) *A right circular cylinder, radius a, whose centre of gravity is at a
distance c from the axis, is placed in the angle between a horizontal and a
vertical plane, so that its axis is horizontal. If the planes be of equal rough-*

ness, find in what positions it will roll, and in what positions it will slip, and shew that it will just not be able to slip at all if

$$c = a \frac{\mu + \mu^2}{1 + \mu^2},$$

where μ is the coefficient of friction.   [S.]

Taking a vertical section through the centre C of the cylinder, let A, B be the points of contact with the horizontal and vertical planes. Draw lines APQ, ASR, BRQ, BSP on both sides of the normals AC, BC, making with the normals angles equal to the angle of friction λ.

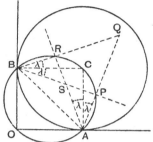

Then since

$CAP = CBP = \lambda$, and $CAR = CBR = \lambda$,

therefore the points P and R lie on the circle ABC, i.e. the circle on AB as diameter, so that

$BPA = BRA = BCA = $ a right angle.

Now there are only three forces acting on the cylinder, its weight and the reactions at A and B, and the latter cannot intersect outside the quadrilateral $PQRS$ if equilibrium is maintained. But this alone is not a sufficient test, for it must also be possible for the reactions to be directed so that they can have a vertical resultant.

Suppose that the line of action of the weight lies between B and R in the figure. It has a counter-clockwise moment about C, but the cylinder cannot roll in this sense, nor is there enough friction to maintain equilibrium, therefore the cylinder slides.

If the cylinder is placed so that the line of action of the weight passes between R and C, it will be in equilibrium, because the reactions at A and B can intersect on the line of action of the weight and balance it. But if the line of action of the weight falls to the right of C, it is no longer possible for the reactions to balance the weight although they fall within the cones of friction at A and B. There will in fact then be no reaction at B and equilibrium will be broken by the rolling of the cylinder at A.

It follows that slipping will not take place unless the vertical through the centre of gravity lies between B and R, and this will just not be possible if

$$c = a - BR \cos \lambda$$
$$= a - BA \sin\left(\frac{\pi}{4} - \lambda\right) \cos \lambda$$
$$= a - a\left(\cos \lambda - \sin \lambda\right) \cos \lambda$$
$$= a\left(\sin^2 \lambda + \sin \lambda \cos \lambda\right)$$
$$= \frac{a\left(\mu^2 + \mu\right)}{\mu^2 + 1},$$

where $\mu = \tan \lambda.$

(iii) *A uniform circular cylindrical log of radius a and weight W lies with its axis horizontal between two rough parallel horizontal rails at the same level and at a distance 2a sin α apart; shew that, if a gradually increasing couple be applied to the log in a plane perpendicular to the rails and axis, the log will turn over one of the rails when the couple is of magnitude Wa sin α, provided the angle of friction λ is greater than α; but otherwise the log will turn about the axis when the couple is*

$$Wa \sin \lambda \cos \lambda \sec \alpha. \qquad [\text{S.}]$$

Taking a vertical section through the centre of the cylinder at right angles to its axis, let C be the centre and A, B points of contact of the cylinder with the rails. Then

$AB = 2a \sin \alpha$, so that the angle $ACB = 2\alpha$.

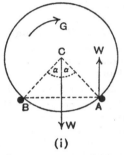

(i)

Assuming that equilibrium is about to be broken by turning about the rail at A, there is then no reaction at B and the cylinder is acted upon by the couple G, its weight W and the reaction at A, and the latter must therefore be vertical and equal to W, forming with the weight a couple $Wa \sin \alpha$, so that when G exceeds $Wa \sin \alpha$ the cylinder will turn over the rail, but only provided that the reaction at A falls within the cone of friction at A. The axis of the cone is the normal AC and the reaction at A is W making an angle α with AC, therefore the condition is satisfied if $\lambda > \alpha$.

If $\lambda < \alpha$, the reaction at A, as determined above, falls outside the cone of friction, so that the assumption of turning about the rail at A is not justified, and we must assume that the cylinder remains in contact with both rails.

There are now two methods of completing the solution. We may draw the resultant reactions at A and B inclined to the normals at the angle of friction λ and meeting at O. The resultant of these reactions together with the weight W must form a couple which, in limiting equilibrium, balances the couple G.

Therefore, in fig. (ii),

$$G = W \cdot MN = \tfrac{1}{2} W (BN - NA)$$
$$= \tfrac{1}{2} W \{BO \sin(\alpha + \lambda) - OA \sin(\alpha - \lambda)\}.$$

But O lies on the circle ABC, since the angle $OBC = \lambda = OAC$, and the diameter of this circle is

$$AB/\sin ACB = a \sec \alpha;$$

therefore $BO = a \sec \alpha \cos(\alpha - \lambda)$ and $OA = a \sec \alpha \cos(\alpha + \lambda)$,

and $\quad G = \tfrac{1}{2} Wa \sec \alpha \{\cos(\alpha - \lambda) \sin(\alpha + \lambda) - \cos(\alpha + \lambda) \sin(\alpha - \lambda)\}$
$\quad\quad = Wa \sec \alpha \sin \lambda \cos \lambda.$

The alternative method of procedure is to put in the normal components of reaction R and S at A and B and the frictions μR and μS, as in fig. (iii).

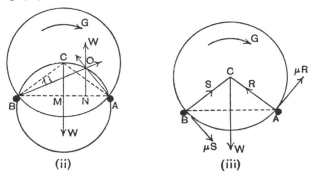

(ii)   (iii)

Then the couple which, in limiting equilibrium, balances the couple G is found by taking the sum of the moments of all the other forces about any point in the plane. Taking moments about C we get

$$G = \mu (R + S)\, a.$$

But by resolving horizontally and vertically, we get

$$\mu (R + S) \cos \alpha = (R - S) \sin \alpha,$$

and $$\mu (R - S) \sin \alpha + (R + S) \cos \alpha = W.$$

Therefore $$(1 + \mu^2)(R + S) \cos \alpha = W,$$

and $$G = \frac{\mu a}{1 + \mu^2}\, W \sec \alpha,$$

or, since $\mu = \tan \lambda$, $$G = W a \sec \alpha \cos \lambda \sin \lambda.$$

9·4. Friction in Unknown Directions.

A different kind of problem from that discussed so far is one in which the directions in which the different points of a body begin to move when equilibrium is broken are not known *a priori* but have to be determined. The directions in which the frictions act are therefore also unknown. The method of procedure is to assume some position for the instantaneous centre of rotation* of the body which is about to move. The direction of motion of each point is then at right angles to the line joining it to the instantaneous centre and the frictions therefore also act at right angles to the lines joining points to the instantaneous centre. We then have the usual three equations of equilibrium and,

* v. *Dynamics*, 5·41.

when the problem is a determinate one, these serve to determine the position of the instantaneous centre of rotation and a relation between the forces.

9·41. Examples. (i) *A uniform heavy rod AB lying on a rough table has a force applied at the end A which is gradually increased until the rod is just about to move in such a way that the initial motion will be rotation about a point C in AB. Prove that the force at A must be applied at right angles to AB and that the point C must divide AB so that*

$$AC : CB = 1 + \sqrt{2} : 1. \qquad \text{[S.]}$$

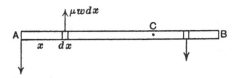

Suppose that the rod is just about to rotate about C in the counterclockwise sense. Then the friction at every point must act so as to have a clockwise moment about C. If dx denotes the length of an element at a distance x from A, its weight may be denoted by $w\,dx$, so that the vertical reaction of the table upon it is $w\,dx$ and the friction acting upon it is $\mu w\,dx$, where μ is the coefficient of friction. The only forces acting on the rod in the horizontal plane are the frictions and the force applied at A, and since the frictions are all at right angles to AB, therefore in equilibrium the force at A must also be at right angles to AB.

The position of C may then be determined by taking moments about A. Let $AB = l$, and $AC = a$, then noting that the frictions on AC and CB are in opposite senses, we get

$$\int_0^a \mu w x\,dx = \int_a^l \mu w x\,dx,$$

or $$\tfrac{1}{2}\mu w a^2 = \tfrac{1}{2}\mu w (l^2 - a^2),$$

so that $$\sqrt{2}\,a = l.$$

Therefore $$AC : CB = a : l - a$$
$$= 1 : \sqrt{2} - 1$$
$$= \sqrt{2} + 1 : 1.$$

We observe that the *three* equations of equilibrium are made use of as follows:

(α) a resolution in the direction AB leads to the fact that the force applied at A has no component along the rod;

(β) the equation of moments about A determine the position of the instantaneous centre C;

(γ) a resolution at right angles to the rod will now determine the least force at A that will just move the rod, say

$$P = \int_0^a \mu w\, dx - \int_a^l \mu w\, dx$$
$$= \mu w (2a - l) = \mu w l (\sqrt{2} - 1).$$

(ii) *A uniform rod AB is lying in a horizontal position upon two parallel horizontal equally rough rods and is at right angles to them. The points of contact are C and D, C being nearest to A. A gradually increasing force is applied to the rod AB at A parallel to the other two rods. Shew that if $\dfrac{4}{AB} < \dfrac{1}{AC} + \dfrac{1}{AD}$ equilibrium will ultimately be broken by the rod turning round D and slipping at C.*   [S.]

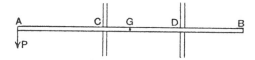

Let W be the weight of the rod AB acting at its middle point G. Then by taking moments about C or D we find that the forces of support at C and D are $W \dfrac{GD}{CD}$ and $W \dfrac{CG}{CD}$ respectively.

If the rod were to turn about any point but C or D the maximum amount of friction would be called into play at both C and D, and if it is possible for there to be a state of limiting equilibrium in which an amount of friction smaller than the maximum is called into play at C or at D this will be the actual state.

Suppose then that when the force at A is P, the friction at C is $k \cdot W \dfrac{GD}{CD}$ and that at D is $k' \cdot W \dfrac{CG}{CD}$, these forces being at right angles to AB, the former opposed to P and the latter in the same sense as P in the equilibrium state.

By moments about A we have

$$k \cdot W \frac{GD}{CD} AC = k' \cdot W \frac{CG}{CD} AD \quad \ldots\ldots\ldots\ldots\ldots(1),$$

and, by resolving at right angles to AB, we have

$$k \cdot W \frac{GD}{CD} - k' \cdot W \frac{CG}{CD} = P \quad \ldots\ldots\ldots\ldots\ldots(2).$$

It follows from (1) that k and k' are always in a definite ratio and then from (2) that k and k' increase with P. Hence, if μ be the coefficient of friction, as P increases, equilibrium will be maintained until k or k' reaches the value μ, and if k is the first to become equal to μ, equilibrium will be broken by sliding at C and turning at D. This will be so if $k > k'$,

i.e. if $\qquad\qquad CG.AD > GD.AC$,

or if $\qquad\quad (\tfrac{1}{2}AB - AC)AD > (AD - \tfrac{1}{2}AB)AC$,

or if $\qquad\qquad\qquad \dfrac{4}{AB} < \dfrac{1}{AC} + \dfrac{1}{AD}.$

(iii) *A uniform triangular lamina rests on a rough horizontal plane, supported upon short pegs of equal length at its corners. To find the least couple which will cause the lamina to turn in its plane.*

It is easy to shew that three equal parallel forces acting at the corners of a triangle have a resultant passing through the centre of gravity, so that if W be the weight of the triangle the pressures on the pegs will each be $\tfrac{1}{3}W$.

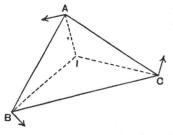

In limiting equilibrium the forces due to friction just balance the required couple, and their directions are at right angles to the lines joining the corners A, B, C to the instantaneous centre of rotation.

Hence if there is an instantaneous centre I inside the triangle, so that slipping takes place simultaneously at A, B and C, the frictions are all limiting and equal to $\tfrac{1}{3}\mu W$ acting at right angles to IA, IB, IC. Since these forces are equivalent to a couple, they would balance one another when turned through a right angle to act along IA, IB, IC; but they are equal forces, so by Lami's Theorem the angles AIB, BIC, CIA are equal. Hence the position of I can be found by drawing on AB and BC segments of circles containing angles of $120°$. The couple required is then the sum of the moments about I of the frictions, i.e.

$$\tfrac{1}{3}\mu W\,(IA + IB + IC).$$

It is easy to see that the frictions could not be equivalent to a couple if the instantaneous centre were outside the triangle, so that the above solution would not hold good if the triangle had an angle greater than $120°$.

To consider the possibility of the triangle turning about one corner, say the corner A. The frictions at B and C would then each be $\tfrac{1}{3}\mu W$, and that at A should be less than the limiting friction, i.e. less than $\tfrac{1}{3}\mu W$, and the three should be equivalent to a couple; so that, as above, when turned through a right angle they should be three forces acting at A in equilibrium. But the resultant of $\tfrac{1}{3}\mu W$ along AB and $\tfrac{1}{3}\mu W$ along AC is $\tfrac{2}{3}\mu W \cos \tfrac{1}{2}A$ and this can only be in equilibrium with a force less than $\tfrac{1}{3}\mu W$, if $\cos \tfrac{1}{2}A$ is less than $\tfrac{1}{2}$; i.e. if $A > 120°$.

It follows that the lamina will only turn about a corner if the angle at that corner is greater than $120°$, and if this be the angle at A, the least couple required is $\quad \tfrac{1}{3}\mu W\,(AB + AC).$

9·5. Miscellaneous Problems. (i) *A cart-wheel.*

Consider a wheel which turns upon an axle whose circular section is only slightly smaller than the aperture in the wheel, so that the wheel and axle may be regarded as having a common centre C.

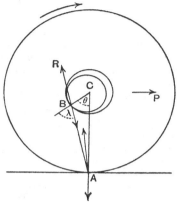

Let a be the radius of the wheel and b that of the axle.

Let A be the point of contact of the wheel with the ground and suppose the wheel about to move from left to right in the figure. There are only three forces acting upon the wheel:

(α) its weight acting along CA;

(β) the reaction of the ground at A, acting on the left of the normal AC, because the friction opposes the tendency to slide at A. (The friction at A is *not* limiting friction.)

(γ) The third force is the pressure of the axle on the wheel, and since this force must balance the other two acting at A, therefore it acts along BA, where B is the point of contact of the wheel and axle, and B must be on the left of CA.

Also since the friction between the wheel and axle is limiting friction, therefore BA makes an angle with the normal CB equal to the angle of friction λ. Then the angle θ between BC and CA is given by

$$\sin(\lambda - \theta) = \frac{b}{a}\sin\lambda.$$

Also, if P be the horizontal force applied to the axle in the direction of motion, R the reaction between the wheel and axle and W the load, then
$$P = R\sin(\lambda - \theta) \quad \text{and} \quad W = R\cos(\lambda - \theta).$$

(ii) *A window sash with a broken cord.*

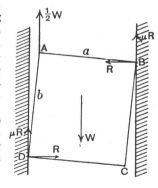

Let $ABCD$ represent a loosely fitting window sash of breadth a and height b. The cord at B is broken and that at A passes over a pulley and carries a counterpoise of weight $\frac{1}{4}W$, where W is the weight of the sash. It is required to find the least coefficient of friction μ that will prevent the sash from sliding down.

Since the sash is loosely fitting there will be contact with the frame at opposite corners B and D only, and by resolving horizontally we find that the reactions there must be equal and opposite, of magnitude R say.

The friction forces μR act vertically upwards at B and D. Then neglecting the slight deviations of the sides of the sash from the vertical and taking moments about B, we get

$$\tfrac{1}{2}Wa - W\tfrac{1}{2}a + \mu Ra - Rb = 0,$$

so that $\qquad\qquad\qquad \mu = b/a.$

(iii)　*Braking of a carriage on an inclined plane.*

A carriage stands on an inclined plane with its axles horizontal. To determine which pair of wheels should be locked in order to admit of the greatest possible gradient for a given coefficient of friction.

Let G be the centre of gravity of the carriage. Assuming the carriage to be symmetrical about the vertical plane through G at right angles to the inclined plane, we can treat the problem as a plane one. Let A, B be points of contact of the wheels with the inclined plane, let the vertical through G cut the plane in C and let α be the inclination of the plane to the horizontal.

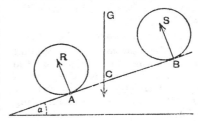

Let R and S be the normal reactions at A and B. In limiting equilibrium the friction between the locked wheels and the plane will be μ times the normal reaction, and the friction between the wheels that are free to rotate and the plane will be negligible in comparison. Hence the friction is μR or μS according as the wheels at A or B are locked.

By taking moments about C for the equilibrium of the carriage, we get

$$R \cdot AC = S \cdot CB.$$

Hence R is greater or less than S according as AC is less or greater than CB.

But the friction μR or μS balances the resolved part of the weight down the plane, viz. $W \sin \alpha$, where W is the weight of the carriage. Hence the greatest possible gradient for a given μ corresponds to the greater value of R or S. Therefore the wheels at A or at B should be locked according as AC or CB is the smaller.

(iv)　*A pair of compasses.*

A pair of compasses is used to describe circles on a horizontal piece of paper, and has legs (of equal length l) always in a vertical plane. The handle at the joint is always vertical. If ϕ be the angle of friction between the compass pencil and the paper, 2α the angle between the legs, and W the vertical pressure on the handle, shew that a horizontal couple $Wl \sin \alpha \tan \phi$ must be applied to the handle in drawing a circle, the joint being supposed to be clamped.

When the joint is clamped, the angle 2α at the joint is capable of being increased elastically by a small amount kB when a bending moment B is applied to the joint in the plane of the legs. When a circle is about to be *drawn as above, it is found that this increase in* 2α *takes place as the weight* W *is applied, provided* $\alpha > \phi$. *If* $\alpha < \phi$, *the increase takes place just after the start. Shew that, as a result, the radius of the circle drawn in the latter case is increased by* $\frac{1}{4}kWl^2\sin 2\alpha$.   [T.]

Let A be the joint, O the centre of the circle and P the point of the pencil. By taking moments about O in the vertical plane we find that the vertical reaction at P is $\frac{1}{2}W$. It follows that when P is describing the circle the friction is $\frac{1}{2}W\tan\phi$ along the tangent to the circle, and since the radius of the circle is $2l\sin\alpha$, therefore, by moments about O, the horizontal couple necessary to overcome the friction is $Wl\sin\alpha\tan\phi$.

Next consider the case in which the joint can yield elastically to a bending moment. Suppose that a vertical pressure W is applied to the handle; then the vertical reaction at P is $\frac{1}{2}W$ as above, and before any horizontal couple is applied the joint tends to yield and the pencil point P tends to slip outwards along OP.

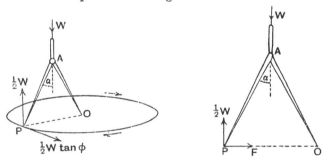

The friction F is therefore along PO and, by taking moments about A for the leg AP, we get

$$Fl\cos\alpha = \tfrac{1}{2}Wl\sin\alpha, \quad \text{or} \quad F = \tfrac{1}{2}W\tan\alpha.$$

This is the amount of friction necessary to maintain equilibrium; but the maximum amount available is $\frac{1}{2}W\tan\phi$, since ϕ is the angle of friction; therefore, if $\alpha > \phi$, equilibrium will be broken by the pencil point slipping outwards along OP while the pressure W is being applied.

Next suppose that $\phi > \alpha$; then sufficient friction will be called into play to prevent the pencil point from slipping along OP, and no motion will take place until a horizontal couple is applied to make P move at right angles to OP. As soon as this motion begins the friction ceases to act along PO and acts at right angles to PO opposing the motion. The only force at P in the plane OAP of the legs is now the vertical reaction $\frac{1}{2}W$ and this produces at A a bending moment

$$B = \tfrac{1}{2}Wl\sin\alpha.$$

The joint then yields so that the angle OAP becomes $2\alpha + kB$, and the radius of the circle becomes

$$2l\sin(\alpha + \tfrac{1}{2}kB).$$

Expanding this sine and neglecting higher powers of k than the first, we get for the radius of the circle

$$2l \sin \alpha + 2l \cdot \tfrac{1}{2} k B \cos \alpha;$$

so that the increase in the radius is

$$lkB \cos \alpha \quad \text{or} \quad \tfrac{1}{4} k W l^2 \sin 2\alpha.$$

EXAMPLES

1. A particle rests on a rough horizontal plane under the action of a horizontal force P. A gradually increasing horizontal force is applied to it perpendicular to P. Determine the direction in which the particle will begin to move. [S.]

2. Find the magnitude and direction of application of the least force necessary to drag a particle up a rough inclined plane. [S.]

3. Prove that the least force, which applied to a uniform heavy sphere of weight W will maintain it in equilibrium against a rough vertical wall, is $W \cos \lambda$ provided λ the angle of friction is less than

$$\cos^{-1} \frac{\sqrt{5}-1}{2}. \qquad \text{[S.]}$$

4. A circular hoop hangs over a horizontal peg and a weight 1·5 times that of the hoop hangs tangentially from it. Shew that if the hoop is about to slip on the peg the coefficient of friction is ·75. [I.]

5. A solid hemisphere of weight W rests in limiting equilibrium with its curved surface on a rough inclined plane, and its plane face is kept horizontal by a weight P attached to a point in its rim. Prove that the coefficient of friction is

$$\frac{P}{\sqrt{W(2P+W)}}. \qquad \text{[S.]}$$

6. A heavy uniform rectangular block whose base is a square of side $2a$ and whose height is $2b$ stands on a rough horizontal plane of coefficient μ. A horizontal force is applied to the middle point of the top of one of the vertical faces and is increased until equilibrium gives way; shew that this happens by the block tilting about an edge of the base or by slipping according as μ is $>$ or $< \dfrac{a}{2b}$. [S.]

7. A square is placed with its plane vertical and one side resting along the line of greatest slope on a rough inclined plane, the coefficient of friction being μ, a string is attached to the upper corner and pulled in a direction parallel to the line of greatest slope up the plane; shew that if the tension be gradually increased the square will slide or tilt according as the angle of inclination of the plane is less or greater than $\tan^{-1}(1-2\mu)$. [S.]

8. A sphere rests on a rough inclined plane, the coefficient of friction being μ, and is supported by a horizontal string attached to the highest point of the sphere and to the plane. Find the inclination of the plane when the sphere is on the point of slipping. [S.]

9. Two weights, W_1 and W_2, rest on a rough plane inclined at an angle α to the horizon, being connected by a string which lies along a line of greatest slope. If μ_1, μ_2 are their coefficients of friction with the plane, and $\mu_1 > \tan \alpha > \mu_2$, prove that, if they are both on the point of slipping,

$$\tan \alpha = \frac{\mu_1 W_1 + \mu_2 W_2}{W_1 + W_2}.$$ [S.]

10. Spheres whose weights are W, W' rest on two inclined planes. The highest points of the spheres are connected by a horizontal string perpendicular to the common horizontal edge of the two planes and above it. If the coefficients of friction μ, μ' are such that each sphere is on the point of slipping down, then

$$\mu W = \mu' W'.$$ [S.]

11. A uniform straight rod rests in a vertical plane with one end resting against a rough vertical wall and the lower end on a rough horizontal plane. If the friction is limiting at both ends when the inclination to the horizontal is α, and the coefficient of friction is the same for both contacts, prove that the angle of friction is $\frac{\pi}{4} - \frac{\alpha}{2}$. [S.]

12. A heavy circular hoop is hung over a rough peg. A weight equal to that of the hoop is attached to it at a given point. Find the coefficient of friction between the peg and the hoop so that the system may hang in equilibrium whatever point of the hoop is placed in contact with the peg. [S.]

13. A cylindrical hole of radius a is bored through a body and the body is suspended from a rough horizontal peg passing through the hole. Prove that in equilibrium the inclination to the vertical of the plane through the centre of gravity of the body and the axis of the hole is not greater than $\sin^{-1} \frac{a \sin \lambda}{d}$, where λ is the angle of friction and d is the distance between the centre of gravity and the axis of the cylinder.

14. A hoop of mass M hangs from a rough peg in a vertical plane. An insect of mass m starts from the lowest point of the hoop and crawls slowly upwards. Prove that the insect can reach the peg if $\sin \lambda > \frac{m}{M+m}$; but that the hoop will slip on the peg when the insect has traversed an arcual distance $a \left[\lambda + \sin^{-1} \left(\frac{M+m}{m} \sin \lambda \right) \right]$ if $\sin \lambda < \frac{m}{M+m}$. λ is the angle of friction between the hoop and the peg, and a is the radius of the hoop. [S.]

15. A semicircle, of radius a, is fixed with its diameter on the ground and its plane vertical. In the same vertical plane a uniform heavy rod, of length l, rests touching the semicircle with one end on the ground. The coefficient of friction between the rod and both the ground and the semicircle being $\tan^{-1}\lambda$, shew that θ, the inclination of the rod to the horizon in the limiting position of equilibrium, is given by

$$l\sin^2\theta = a\sin 2\lambda.$$ [S.]

16. A uniform rod of weight W and length $2b$ carries a weight P at its upper end and passes over one peg and under another, the pegs being at a distance a and the line joining them making an angle α with the horizontal. Prove that when sliding is just about to take place, the distance of the upper end of the rod from the point midway between the pegs is $\dfrac{Wb}{W+P}+\dfrac{a\tan\alpha}{2\mu}$ (μ being the coefficient of friction). [S.]

17. A string of length l has two light rings fastened at its ends, these slide upon a rough straight wire fixed at an inclination β to the horizon; a heavy smooth ring has the string passing through it, and hangs in equilibrium; shew that the greatest possible distance between the rings on the wire is $\dfrac{l(\mu-\tan\beta)}{(1+\mu^2)^{\frac{1}{2}}}$, μ being the coefficient of friction between the rings and the wire, and being greater than $\tan\beta$. [S.]

18. A uniform rod AB, whose weight can be neglected, is placed with its middle point C at the highest point of a fixed rough horizontal circular cylinder, the rod being at right angles to the axis of the cylinder. Two weights W, W' are now suspended from the ends A, B, and the rod is rotated on the cylinder, without sliding, until it is in a new position of equilibrium. Prove that this is impossible unless the angle of friction is greater than $\dfrac{l}{a}\dfrac{W\sim W'}{W+W'}$, where $2l$ is the length of the rod and a the radius of the cylinder. [I.]

19. A uniform rod of length $2l$ rests within a hollow sphere of radius a in a vertical plane through the centre of the sphere. The sphere is rough, the angle of friction being λ. Shew that, if $l < a\cos\lambda$, the greatest inclination to the horizontal at which the rod can rest is given by the equation
$$(2l^2-a^2)\sin\theta - a^2\sin(\theta-2\lambda) = 0.$$
How can the rod rest if $l > a\cos\lambda$? [S.]

20. A heavy beam inclined at an angle α to the horizontal rests with one end against a vertical wall and the other on the ground, the coefficient of friction in each case being $\tan\lambda$. The beam is in a vertical plane perpendicular to the plane of the wall. Shew that if the beam is kept from slipping down by a horizontal string of tension T attached to the lower end, *or* by a vertical string of tension T' attached to the upper end, then
$$T' = T\tan(\alpha+\lambda).$$ [S.]

21. A rough chain lies across the ridge of a double inclined plane, stretching down the line of greatest slope on either side of the ridge. Prove that it will be in limiting equilibrium when the line joining its two ends is inclined to the horizon at the angle of friction. [C.]

22. Two beads A, B, whose weights are w_1, w_2, are tied to the ends of a string, on which is threaded a third bead C of weight W. The beads A, B can slide on a rough horizontal rod, whose coefficient of friction with each bead is μ. If, when A, B are as far apart as possible, the strings AC, BC each make an angle θ with the vertical, prove that, if $w_1 > w_2$,
$$\tan \theta = \mu \{1 + 2\,(w_2/W)\}.$$ [S.]

23. Three rods AB, BC, CD, of equal length (l) and weight, are loosely jointed together at B and C. Two light rings are fastened at A and D and slide on a rough horizontal rod AD. If μ is the coefficient of limiting friction between the rings and the rod, shew that the greatest length of AD is
$$l\left(1 + \frac{6\mu}{\sqrt{9\mu^2 + 4}}\right).$$ [C.]

24. A uniform rod rests within a fixed vertical circle, subtending an angle 2α at the centre; its upper end is smooth, and its lower end rough (coefficient of friction $= \tan \lambda$); shew that the angle which the rod makes with the horizon cannot be greater than θ, where
$$\tan \theta = \frac{\sin \lambda}{\cos \lambda + \cos (\lambda + 2\alpha)}.$$ [S.]

25. A uniform solid rectangular block leans against a rough vertical wall, with one edge resting on a rough horizontal plane. If θ is the angle which the plane containing the two supporting edges makes with the vertical, and λ is the angle of friction for both edges, prove that, in order that the friction may be limiting at both edges, the thickness of the block measured perpendicular to this plane must bear to the distance between the edges the ratio of
$$\tan \theta - \tan 2\lambda \quad \text{to} \quad \sec 2\lambda.$$ [S.]

26. A uniform heavy elliptic lamina rests with its minor axis ($2b$) vertical on a rough horizontal plane. A string is attached to the centre and is pulled horizontally in the plane of the lamina until the major axis ($2a$) of the lamina is vertical. Shew that if there is no slipping the coefficient of friction between the lamina and the horizontal plane cannot be less than $(a^2 - b^2)/2ab$. [S.]

27. A cube of weight W and edge $2a$ rests on a rough horizontal plane. A ladder of weight W' and length $2l$ rests against a smooth vertical face of the cube, making an angle α with the plane. If the cube were fixed, a man of weight w could ascend the ladder to a vertical

height h before it slips. Prove that, if he attempts to ascend, the cube will turn round its further edge before he reaches this height, if

$$h > (Wa - W'l\cos\alpha)\tan\alpha/w;$$

the coefficient of friction between the cube and the ground being $> a/2l\sin\alpha$, and $2l\sin\alpha > a$. [S.]

28. A uniform rod AB 24 in. in length is lying in a horizontal position upon two parallel horizontal equally rough rods and at right angles to them. The points of contact are C, D. AC is 11 in., AD 18 in. A gradually increasing force is applied at A to the rod AB parallel to the other two rods. Shew that the equilibrium will ultimately be broken by the rod turning about C and slipping at D. [C.]

29. A uniform rod AB of weight W and length l lies on a horizontal plane whose coefficient of friction is μ. A string is attached to B and is pulled in a horizontal direction perpendicular to the rod. As the tension is gradually increased, shew that the rod begins to turn about a point in it whose distance from A is approximately $\frac{3l}{10}$, and the tension of the string is then about $\frac{2\mu W}{5}$. [S.]

30. A uniform rod AB of weight W rests horizontally on two equally rough supports at A and C. Prove that the least horizontal force applied at B in a direction perpendicular to BA which is able to move the rod is $\frac{1}{2}\mu W$ or $\mu W\frac{b-a}{2a-b}$, according as $3b$ is greater or less than $4a$, where $AB = 2a$, $AC = b$, and μ is the coefficient of friction. [S.]

31. A uniform rod of length $2a$ rests on a rough plane whose inclination to the horizon (α) is greater than the angle of friction (λ) and can turn about a point in it distant b from the centre of gravity of the rod.

Prove that for equilibrium the greatest possible inclination of the rod to the line of greatest slope is

$$\sin^{-1}\frac{\tan\lambda}{\tan\alpha}\left(\frac{b^2+a^2}{2ab}\right).$$ [S.]

32. A uniform square lamina $ABCD$ is supported by three small pegs rigidly attached to the lamina at the corners B, C and the middle point of the side AD, the pegs resting on a rough horizontal plane. Find the least force in the line AD that will move the lamina and find the point about which it begins to turn. [S.]

33. A solid cylinder of weight w and of radius R rests with its axis vertical on a rough horizontal plane. If the coefficient of friction between the surfaces in contact be μ, shew that the couple required to rotate the cylinder about its axis is $\frac{2}{3}\mu wR$, assuming that the normal pressure on the base of the cylinder is uniformly distributed over the area of the base. Find also what couple will be required if it be assumed that the normal pressure per unit area of the base varies inversely as the distance from the axis of the cylinder. [S.]

34. A uniform rod of mass M is placed horizontally on a rough inclined plane of angle $\alpha\,(<\tan^{-1}\mu)$. A string fastened to one end is pulled downwards in the direction of the line of greatest slope, the pull being slowly increased until the rod begins to move. Find the point about which it will commence to turn, and shew that in that case the tension of the string

$$= \lambda\,(\sqrt{2}-\lambda)\,\mu Mg \cos\alpha,$$

where $\lambda = \sqrt{1+\tan\alpha/\mu}$. [I.]

35. Two equal heavy uniform circular cylinders lie on a rough horizontal plane and touch along a generator. A third equal rough cylinder is placed on the top of the former and touching each along a generator. Shew that in order that equilibrium may be possible in this configuration the angle of friction at any point of contact cannot be less than $15°$. There is no action between the two lower cylinders. [I.]

36. Two rough planes are equally inclined at an angle α to the horizontal. A cylinder of radius a, whose centre of mass is at distance c from its axis, rests between them. If λ be the angle of friction between the cylinder and each plane $(\lambda < \frac{1}{2}\pi - \alpha)$, shew that it is impossible for the cylinder to be placed in any position of limiting equilibrium if

$$c < a \sin\alpha \sin 2\lambda/\sin 2\alpha. \qquad\text{[S.]}$$

37. Two rough planes which intersect in a horizontal straight line are inclined at angles α and β to the horizon. A cylindrical ruler whose length is $2a$ and diameter $2b$ rests with one end on each plane, the axis of the ruler being at right angles to the line of intersection of the planes. Shew that, if the ruler be in limiting equilibrium with its axis inclined at an angle θ to the horizon,

$$\cos(\theta+\phi)\cot(\alpha+\lambda) - \cos(\theta-\phi)\cot(\beta-\lambda) = 2\sin\theta\cos\phi,$$

where $\tan\lambda$ is the coefficient of friction for each end and $\cot\phi = a/b$. [S.]

38. Two equal uniform rods AB, BC are freely hinged at B, C rests on a rough horizontal plane and A is attached to a point above it. When C is as far as possible from A for equilibrium, AB, BC make angles α, β with the vertical. Prove that the coefficient of friction between the rod at C and the plane is $2/(\tan\alpha - 3\tan\beta)$. [S.

39. Two uniform beams AB, AC of the same length are smoothly hinged together at A and placed standing in a vertical plane with the ends B, C on a rough horizontal plane. Shew that, if the weight of one is double that of the other, the least value of the coefficient of friction necessary for equilibrium is $\frac{2}{3}\tan\frac{1}{2}(BAC)$.

If the coefficient of friction has this value, at which end is the friction limiting? [S.]

40. Two ladders of equal length but unequal weights, hinged together, form a step-ladder, the weights of the two parts being W_1 and W_2 respectively; and equilibrium is maintained by friction between the

ladders and the ground, the coefficient of friction being μ. Find the inclination of the ladders at which slipping is on the point of taking place; and calculate the action at the hinge when this is the case. Assume the centre of gravity of each ladder to be at its middle point.

[S.]

41. Two equal uniform ladders are jointed at one end and stand with the other ends on a rough horizontal plane. A man whose weight is equal to that of one of the ladders ascends one of them. Prove that the other will slip first.

If it begins to slip when he has ascended a distance x, prove that the coefficient of friction is $(a+x)\tan\alpha/(2a+x)$, a being the length of each ladder, and α the angle each makes with the vertical. [S.]

42. A uniform sphere of weight W rests on a horizontal plane touching it at C. A uniform beam AB of weight X has its end A on the plane and is a tangent to the sphere at B, ABC being a vertical plane. The ratios of the tangential to the normal reactions at A, B, C in equilibrium are α, β, γ. Shew that β is greater than α and than γ, and that $\alpha > \gamma$ if $W > X(1 - \cos BAC)$. [S.]

43. On the top of a fixed rough sphere of radius r rests a thin uniform plank, and a man stands on the plank at this point. Shew that he can walk slowly a distance $(n+1)r\epsilon$ along the plank without its slipping off the sphere, if the weight of the plank is n times that of the man, and ϵ is the angle of friction between the plank and cylinder. [S.]

44. A railway truck is at rest on an incline of slope α with the lower pair of wheels locked. Shew that the coefficient of friction μ between the wheels and the rails must not be less than $(a+b)/(h+b\cot\alpha)$, where h is the distance of the centre of gravity of the truck from the plane of the rails, and a, b are the distances of the centre of gravity from the lower and upper axles measured parallel to the incline. [S.]

45. A light string, supporting two weights w and w', is placed over a wheel (radius a) which can turn round a fixed rough axle (radius b, friction coefficient μ). There being no slipping of the string on the wheel, shew that the wheel will just begin to rotate round the axle if

$$(w-w')\,a = (w+w'+W)\,b\sin\lambda,$$

where $\mu = \tan\lambda$ and W is the weight of the wheel. [S.]

46. A circular axle, of radius a, fits in a bearing whose radius is slightly larger than a; shew that when the axle is just about to turn in the bearing, the reaction between the axle and bearing will touch a concentric circle of radius $a\sin\lambda$, where λ is the angle of friction.

A uniform bar PQ, of weight W, rests in a horizontal position, being free to turn about a horizontal axle, like the above, whose centre is at a distance b from P, the nearer end of the bar. At P is hung a weight W'

which would maintain equilibrium in the absence of friction; prove that if an additional weight X just starts P to move downwards

$$X = (W + W')\,(a\sin\lambda)/(b - a\sin\lambda).$$ [S.]

47. A tripod, formed of three equal uniform rods freely hinged together at one end, rests with its feet at equal distances apart on a rough horizontal plane for which the coefficient of friction is μ. If the foot of each rod is on the point of slipping, find the inclination of each rod to the vertical. [S.]

48. The lower ends of three equal uniform rods, each of weight W, rest upon the ground at the angular points of an equilateral triangle, and their upper ends are loosely tied together at a point, from which also a weight W' is suspended. Shew that the coefficient of friction of the ground must not be less than $\dfrac{1}{2}\cdot\dfrac{3W + 2W'}{3W + W'}\cdot\tan\theta$, where θ is the inclination of each rod to the vertical. [S.]

49. Three uniform rods, VA, VB, VC, of total weight W, are connected at V by a smooth weightless joint, and rest in the form of a tripod on a rough horizontal table with their other ends at A, B, C, points on the table. The projection of V on the table is O, and the angles BOC, COA, AOB are respectively α, β, γ; also the rods make angles θ, ϕ, ψ respectively with the vertical. Prove that the frictions F_1, F_2, F_3 at A, B, C respectively are given by

$$\frac{F_1}{\sin\alpha} = \frac{F_2}{\sin\beta} = \frac{F_3}{\sin\gamma} = \frac{1}{2}\,\frac{W}{\sin\alpha\cot\theta + \sin\beta\cot\phi + \sin\gamma\cot\psi}.$$ [C.]

ANSWERS

1. Making an angle $\cos^{-1}(P/F)$ with P, where F is the limiting friction. 2. $W\sin(\alpha + \lambda)$ making an angle λ with the plane. 8. 2λ. 12. $1/\sqrt{3}$. 19. At any inclination. 32. $\frac{1}{2}\mu W\left(1 + \dfrac{1}{\sqrt{5}}\right)$, where W is the weight. On the line bisecting AD and BC at a distance $\frac{1}{4}AB$ outside BC. 33. $\frac{1}{2}\mu w R$. 39. The end of the lighter beam. 40. $\tan^{-1}\{(W_1 + 3W_2)/(W_1 + W_2)\}$ to the vertical, if $W_1 > W_2$. Horizontal $\frac{1}{4}\mu(W_1 + 3W_2)$, vertical $\frac{1}{4}(W_1 - W_2)$. 47. $\tan^{-1}(2\mu)$.

Chapter X

CENTRES OF GRAVITY

10·1. The *centre of gravity* or *centre of mass* of a body was defined in **4·51** as the centre of the parallel forces which represent the weights of the particles of which the body is composed; and it follows from **4·53** that if m_1, m_2, m_3, ... are numbers proportional to the weights, or to the masses, of a system of particles situated at the points

$$(x_1, y_1, z_1), \quad (x_2, y_2, z_2), \quad (x_3, y_3, z_3), \quad ...,$$

using rectangular axes, then the co-ordinates $\bar{x}$, $\bar{y}$, $\bar{z}$ of the centre of gravity of the particles are given by the formulae

$$\bar{x} = \frac{m_1 x_1 + m_2 x_2 + m_3 x_3 + ...}{m_1 + m_2 + m_3 + ...}, \text{ etc., etc.}$$

or $\qquad \bar{x} = \Sigma mx/\Sigma m, \quad \bar{y} = \Sigma my/\Sigma m, \quad \bar{z} = \Sigma mz/\Sigma m.$

10·11. Further, considering a scattered system of particles of masses $m_1, m_2, m_3, ...$ at points $A_1, A_2, A_3, ...,$ the weights of the particles are proportional to the masses, and on a certain scale can be represented by $m_1, m_2, m_3, ...$ The resultant of the parallel forces m_1 at A_1 and m_2 at A_2 is a force $m_1 + m_2$ acting at a point B_1 on $A_1 A_2$ such that by **4·3**

$$m_1 A_1 B_1 = m_2 B_1 A_2.$$

Then the resultant of $m_1 + m_2$ at B_1 and m_3 at A_3 is $m_1 + m_2 + m_3$ acting at a point B_2 on $B_1 A_3$ such that

$$(m_1 + m_2) B_1 B_2 = m_3 B_2 A_3,$$

and so on.

It follows that what we have defined as the centre of gravity of the system of particles is also what we defined in **2·6** as the centroid of the points $A_1, A_2, A_3, ...$ for multiples $m_1, m_2, m_3, ...$

10·12. The analytical formulae $\bar{x} = \Sigma mx/\Sigma m$, etc., which determine the centre of gravity of a set of particles, may be used in a wider sense to determine the centre of gravity of a set of bodies of finite size, provided that the typical 'm' denotes the mass of such a body and the corresponding x, y, z denote the known co-ordinates of the centre of gravity of that body. This follows from the fact that the grouping together of sets of particles to form finite bodies, as an intermediate step in the process of finding the centre of the parallel forces which represent the weights of *all* the particles, cannot affect the position of that centre.

10·13. It is evident that applications of the standard formulae of **10·1** are possible in which one or more of the m's may be negative. Such a case arises when a body contains a cavity and is such that if the cavity were filled up, the mass and centre of gravity of the whole would be known. We may then find the centre of gravity of the body containing the cavity by considering the weight of the whole body as a positive force from which has to be subtracted the weight of the matter that fills the cavity, and this amounts to treating the m of the cavity as a negative number in the formula.

10·2. Determination of Centres of Gravity. In a large number of cases the positions of centres of gravity can be determined by simple geometrical considerations.

A uniform thin straight rod. Let AB be the rod and G its middle point. The whole rod can be regarded as composed of pairs of equal elements equidistant from G; the centre of gravity of each such pair as at P, Q in the figure is then at the middle point of PQ, i.e. at G. Therefore the centre of gravity of the whole rod is at its middle point.

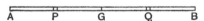

In what follows when we refer to the centre of gravity of a geometrical figure, area or surface, a material sheet of small uniform thickness with the specified boundary is to be understood.

10·21. A parallelogram. Let $ABCD$ be the parallelogram. By drawing a series of lines parallel to BC the whole area can be divided into a series of narrow strips of uniform width such that the centre of gravity of each strip is, by the same argument as in **10·2**, at its middle point. Therefore the centre of gravity of the whole figure lies on the line EF which joins the middle points of BC and 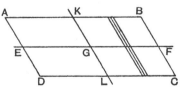 AD, and passes through the middle points of all the strips parallel to BC. By a similar argument, the centre of gravity lies on the line KL which joins the middle points of AB and CD. Hence the intersection of EF and KL is the centre of gravity.

10·22. A triangle. Let ABC be the triangle. The area can be divided into narrow strips by lines parallel to BC. Let bc be any one of these lines, let D be the middle point of BC and let AD cut bc in d. Then since bc is parallel to BC, the triangles Abd and ABD are similar, as are Adc and ADC.

Hence, by using the similarity of the triangles and the fact that $BD = DC$, we have

$$bd : dA = BD : DA = DC : DA = dc : dA,$$

so that $\qquad bd = dc.$

Therefore the centres of gravity of all narrow strips parallel to BC lie on AD (the median). Therefore the centre of gravity of the triangle lies on AD. Similarly, if E is the middle point of AC, the centre of gravity lies on the median BE. Therefore it is at G where AD and BE intersect.

Again, join DE, then since DE is parallel to BA, therefore AGB, DGE are similar triangles, as are ABC and EDC.

Therefore $\qquad\qquad AG : AB = GD : DE,$

or $\qquad\qquad AG : GD = AB : DE = BC : DC$
$$\qquad\qquad\qquad\qquad = 2 : 1.$$

Therefore G is at a point of trisection of the median AD. Similarly

$$BG : GE = 2 : 1.$$

Cor. If F is the middle point of AB, G also lies on CF; but there is only one centre of gravity, therefore the medians AD, BE, CF are concurrent.

10·23. *The centre of gravity of a triangle coincides with that of three particles of the same weight placed at its corners.*

For, if we place particles of weight w at A, B and C, the weights at B and C have a resultant $2w$ at D, and the resultant of $2w$ at D and w at A divides DA at G so that $2DG = GA$, i.e. at the point of trisection which is the centre of gravity of the triangle.

10·24. A quadrilateral. There is no such simple method of constructing the centre of gravity of a quadrilateral as there is for a triangle, but the following theorem is instructive as a method of procedure in a certain class of problems.

The centre of gravity of a quadrilateral coincides with that of four particles of equal mass placed at the four corners together with a fifth particle of equal but negative mass placed at the intersection of the diagonals.

Let $ABCD$ be the quadrilateral and O the intersection of its diagonals.

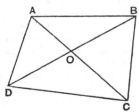

Let m and m' denote the masses of the triangles ABD, CBD. Since these triangles have a common base BD, therefore

$$m : m' = AO : OC.$$

Then by **10·23** above, the triangle ABD may be replaced by particles of mass $\frac{1}{3}m$ at each point A, B, D; and in like manner the triangle CBD by particles of mass $\frac{1}{3}m'$ at each point C, B, D.

Hence the centre of gravity of the quadrilateral coincides with that of four particles: $\frac{1}{3}m$ at A, $\frac{1}{3}(m+m')$ at B, $\frac{1}{3}m'$ at C and $\frac{1}{3}(m+m')$ at D. Further, we may add particles of mass $\frac{1}{3}m'$ at A and $\frac{1}{3}m$ at C, so as to make the four particles at A, B, C, D have the same mass $\frac{1}{3}(m+m')$, if we counterbalance the added masses by placing $-\frac{1}{3}(m+m')$ at the centre of gravity of $\frac{1}{3}m'$ at A and $\frac{1}{3}m$ at C. But, since

$$m : m' = AO : OC,$$

therefore O is this centre of gravity, and we have therefore proved that the centre of gravity of the quadrilateral coincides with that of four particles each of mass $\frac{1}{3}(m+m')$ at A, B, C, D together with a particle of mass $-\frac{1}{3}(m+m')$ at O.

It will be noticed that though we have proved the proposition we have not found the centre of gravity of the quadrilateral.

10·25. A tetrahedron. Let $ABCD$ be the tetrahedron and H the centre of gravity of the base BCD, i.e. the point which divides the median BE so that

$$BH = 2HE.$$

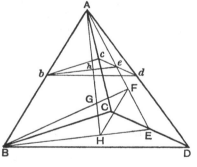

Suppose the tetrahedron to be divided into thin slices by planes parallel to the plane BCD. Let bcd be any one of these sections cutting AE in e and AH in h.

Then since AE is a median of the triangle ACD and cd is parallel to CD, therefore $ce = ed$ and be is a median of the triangle bcd.

Again the parallel planes bcd and BCD are cut by the plane ABE in parallel lines be and BE so that Abh and ABH are similar triangles as are Ahe and AHE.

Therefore
$$bh : hA = BH : HA$$
$$= 2HE : HA$$
$$= 2he : hA,$$

so that $bh = 2he$, and h is the centre of gravity of the triangle bcd. Similarly the centres of gravity of all thin slices parallel to BCD lie on AH, and therefore the centre of gravity of the tetrahedron lies on AH.

By a similar proof, if F is the centre of gravity of the triangle ACD, the centre of gravity of the tetrahedron lies in BF. It is therefore the point G in which AH and BF intersect.

Join HF. Then since

$$BH:HE=2:1=AF:FE,$$

therefore HF is parallel to AB and the triangle AGB, HGF are similar, as are ABE and FHE.

Hence, by similar triangles,

$$AG:AB=GH:FH,$$

or $\qquad AG:GH=AB:FH=BE:HE$

$$=3:1,$$

or $\qquad\qquad GH=\tfrac{1}{4}AH;$

i.e. the centre of gravity of the tetrahedron lies on the line joining the centre of gravity of any face to the opposite vertex and divides it in the ratio $1:3$.

It is easy to verify that the centre of gravity of the tetrahedron coincides with that of four particles of equal mass placed at the corners.

10·26. A pyramid on a plane base. Let V be the vertex and $ABCDE$ the base. By joining one corner of the base to each of the others the base can be divided up into triangles and the pyramid into tetrahedra, and the distance of the centre of gravity of each tetrahedron from the base is one-quarter of the height of the pyramid; therefore the centre of gravity of the pyramid must be at a distance from the base equal to one-quarter of the height of the pyramid.

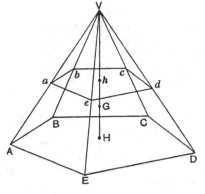

Again the pyramid can be divided into thin slices by planes parallel to the base, making sections, such as $abcde$, which can easily be proved to be similar to $ABCDE$; and if H is the centre of gravity of the base $ABCDE$, it follows from the similarity of the figures as in **10·25** that VH cuts the plane $abcde$ in a point h which is the centre of gravity of the section $abcde$. The centre of gravity G of the pyramid therefore lies on VH and, since its distance from the base of the pyramid is one-quarter of the height, therefore

$$HG=\tfrac{1}{4}HV.$$

10·27. A solid cone with a plane base. A cone may be regarded as a limiting form of a pyramid when the number of sides of its base has been increased and their length decreased without limit so that the base is a continuous curve. Then it follows from **10·26** that, if V is the vertex and H the centre of gravity of the base, the centre of gravity G of the cone lies on VH and is such that

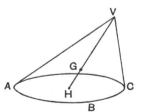

$$HG = \tfrac{1}{4}HV.$$

10·3. Centres of Gravity by Integration. The analytical formulae $\bar{x} = \Sigma mx / \Sigma m$, etc. may be adapted for finding the centre of gravity of a continuous distribution of matter forming a finite body of a given shape. We regard the body as subdivided into a large number of small parts of such shape that the position of the centre of gravity of each part is known to a first approximation. The formulae will then give approximate values to the co-ordinates $\bar{x}$, $\bar{y}$, $\bar{z}$ of the centre of gravity of the body; and the larger the number of subdivisions the more accurate in general will be the result. Unless, however, we can evaluate the sums Σmx, Σmy, Σm, the formulae are useless; but if the number of subdivisions is increased the magnitudes of the parts decreasing without limit the sums Σmx, etc. become definite integrals, which in many cases can be evaluated, and, because in the integrals the small parts tend to vanishing, what was in the first stage approximative now becomes exact. Hence, if we use dm to denote an element of mass of a body and x, y, z are first approximations to the co-ordinates of the centre of gravity of dm, the formulae

$$\bar{x} = \frac{\int x\,dm}{\int dm}, \quad \bar{y} = \frac{\int y\,dm}{\int dm}, \quad \bar{z} = \frac{\int z\,dm}{\int dm},$$

where the integrations are taken through the body, determine its centre of gravity.

The integrations may be along curves, or over surfaces or throughout volumes according to the form of the distribution of matter.

10·31. Curves. If matter be distributed along a plane curve so that ρ is the mass per unit length at a point (x, y), then if ds

denotes an element of arc at (x, y) we may write $dm = \rho\, ds$, and, with sufficient accuracy for our purpose, x, y may be the coordinates of any point of ds, and the centre of gravity of the whole distribution is given by

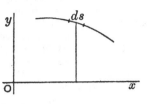

$$\bar{x} = \frac{\int x\rho\, ds}{\int \rho\, ds}, \quad \bar{y} = \frac{\int y\rho\, ds}{\int \rho\, ds}.$$

In the case of a uniform distribution, ρ is constant and

$$\bar{x} = \frac{\int x\, ds}{\int ds}, \quad \bar{y} = \frac{\int y\, ds}{\int ds}.$$

10·311. Circular Arc. Let ACB be a circular arc of radius a which subtends an angle 2α at its centre O.

Take the axis of x along OC, where C is the middle point of the arc, and a perpendicular axis Oy.

If P be any point on the arc and OP makes an angle θ with Ox, the co-ordinates of P are $x = a\cos\theta$ and $y = a\sin\theta$; and, if PP' is a small element of arc, we may write

$$PP' = a\, d\theta.$$

Therefore

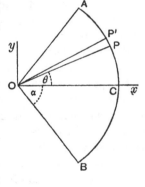

$$\bar{x} = \frac{\displaystyle\int_{-a}^{a} a\cos\theta.a\, d\theta}{\displaystyle\int_{-a}^{a} a\, d\theta} = \frac{a\sin\alpha}{\alpha},$$

and

$$\bar{y} = \frac{\displaystyle\int_{-a}^{a} a\sin\theta.a\, d\theta}{\displaystyle\int_{-a}^{a} a\, d\theta} = 0.$$

Hence the centre of gravity G lies on the middle radius OC (as is obvious by symmetry) and $OG = \dfrac{a\sin\alpha}{\alpha}$.

10·312. Example. *If O is the pole of the lemniscate $r^2 = a^2\cos 2\theta$ and G is the centre of gravity of any arc PQ of the curve, prove that OG bisects the angle POQ.*

Let RR' be an element of arc ds, where R is the point (r, θ).

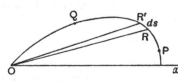

Then, from the equation of the curve, $r\, dr = -a^2\sin 2\theta\, d\theta.$

Also
$$ds^2 = dr^2 + r^2 d\theta^2$$
$$= d\theta^2 \left\{ \frac{a^4 \sin^2 2\theta}{r^2} + r^2 \right\}$$
$$= \frac{a^4 d\theta^2}{r^2}.$$

Then, if $POx = \alpha$ and $QOx = \beta$,

$$\bar{x} = \frac{\int x \, ds}{\int ds} = \frac{\displaystyle\int_\alpha^\beta r \cos\theta . \frac{a^2}{r} d\theta}{\displaystyle\int_\alpha^\beta \frac{a^2}{r} d\theta} = \frac{\sin\beta - \sin\alpha}{\displaystyle\int_\alpha^\beta \frac{d\theta}{r}};$$

and

$$\bar{y} = \frac{\int y \, ds}{\int ds} = \frac{\displaystyle\int_\alpha^\beta r \sin\theta . \frac{a^2}{r} d\theta}{\displaystyle\int_\alpha^\beta \frac{a^2}{r} d\theta} = \frac{\cos\alpha - \cos\beta}{\displaystyle\int_\alpha^\beta \frac{d\theta}{r}}.$$

Therefore $\tan GOx = \dfrac{\bar{y}}{\bar{x}} = \dfrac{\cos\alpha - \cos\beta}{\sin\beta - \sin\alpha} = \tan\tfrac{1}{2}(\alpha + \beta);$

whence $GOx = \tfrac{1}{2}(POx + QOx),$

or OG bisects the angle POQ.

10·32. Areas and Surface Distributions. In the case of matter spread over a surface, if ρ is the mass per unit area at a point (x, y, z) we may write $dm = \rho \, dS$, where dS is an element of area of the surface and the formulae become

$$\bar{x} = \frac{\int x \rho \, dS}{\int \rho \, dS}, \text{ etc.}$$

or, if the density be uniform so that ρ is constant,

$$\bar{x} = \frac{\int x \, dS}{\int dS}, \text{ etc.}$$

The choice of an element dS depends on the nature of the problem.

(i) For example to *find the centre of gravity of a plane area bounded by a curve* $y = f(x)$, *the co-ordinate axes and an ordinate* $x = a$.

If P be a point (x, y) on the curve, a convenient form for dS is a narrow strip of area $y \, dx$, and the co-ordinates of its centre of gravity, with sufficient accuracy for our purpose, are $(x, \tfrac{1}{2}y)$, so that

$$\bar{x} = \frac{\displaystyle\int_0^a xy\,dx}{\displaystyle\int_0^a y\,dx} \quad \text{and} \quad \bar{y} = \frac{\displaystyle\int_0^a \tfrac{1}{2}y^2\,dx}{\displaystyle\int_0^a y\,dx},$$

where in the subjects of integration we make use of the equation of the curve $y = f(x)$.

(ii) *Sectorial area bounded by a curve $r = f(\theta)$ and two radii $\theta = \alpha, \ \theta = \beta$.*

If P be a point (r, θ) on the curve, it is convenient to take for dS a narrow sector POP' whose area is $\tfrac{1}{2}r^2 d\theta$ to the first order of small quantities, and the co-ordinates of its centre of gravity, with sufficient accuracy for our purpose, are

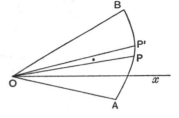

$$x = \tfrac{2}{3}r\cos\theta, \quad y = \tfrac{2}{3}r\sin\theta.$$

Therefore
$$\bar{x} = \frac{\int x\,dS}{\int dS} = \frac{\displaystyle\int_\alpha^\beta \tfrac{2}{3}r\cos\theta \cdot \tfrac{1}{2}r^2 d\theta}{\displaystyle\int_\alpha^\beta \tfrac{1}{2}r^2 d\theta}$$

$$= \frac{2}{3}\frac{\displaystyle\int_\alpha^\beta r^3\cos\theta\,d\theta}{\displaystyle\int_\alpha^\beta r^2\,d\theta},$$

and similarly
$$\bar{y} = \frac{2}{3}\frac{\displaystyle\int_\alpha^\beta r^3\sin\theta\,d\theta}{\displaystyle\int_\alpha^\beta r^2\,d\theta},$$

where in the subjects of integration we make use of the equation of the curve $r = f(\theta)$.

10·321. Circular Sector. Taking a sector of radius a and angle

2α, with the figure of **10·311**, we have $r = a$, so that the results obtained in **10·32** (ii) become

$$\bar{x} = \frac{2}{3} a \frac{\int_{-a}^{a} \cos\theta\, d\theta}{\int_{-a}^{a} d\theta} = \frac{2}{3} a \frac{\sin\alpha}{\alpha}$$

and

$$\bar{y} = \frac{2}{3} a \frac{\int_{-a}^{a} \sin\theta\, d\theta}{\int_{-a}^{a} d\theta} = 0.$$

10·322. Semicircular Area and Quadrant of a Circle. For a semicircle $\alpha = \frac{1}{2}\pi$, so that

$$OG = \frac{4a}{3\pi}.$$

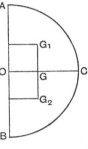

Also the middle radius OC divides the semicircle into two quadrants AOC and BOC whose centres of gravity G_1, G_2 must be such that $G_1 G_2$ passes through G and is parallel to AB. Hence the distance of G_1 from OA is $\frac{4a}{3\pi}$ and by symmetry its distance from OC is the same. So the centre of gravity of a quadrant of a circle of radius a is at a distance $\frac{4a}{3\pi}$ from each of its bounding radii. The same result is of course obtained from putting $\alpha = \frac{1}{4}\pi$ in the result of **10·321**.

10·33. Examples. (i) *Find the co-ordinates of the centre of gravity of a quadrant of the area of the curve $x^{\frac{2}{3}} + y^{\frac{2}{3}} = a^{\frac{2}{3}}$, bounded by the axes of co-ordinates.*

Let OAB be the quadrant whose centre of gravity it is required to find.

Using **10·32** (i) we have

$$\bar{x} = \frac{\int_{0}^{a} xy\, dx}{\int_{0}^{a} y\, dx} \quad \text{and} \quad \bar{y} = \frac{\int_{0}^{a} \frac{1}{2} y^2\, dx}{\int_{0}^{a} y\, dx}.$$

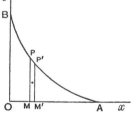

Then, by substituting for y from the equation of the curve,

$$\int_{0}^{a} y\, dx = \int_{0}^{a} (a^{\frac{2}{3}} - x^{\frac{2}{3}})^{\frac{3}{2}}\, dx,$$

and by putting $x = a\sin^3\theta$

$$= 3a^2 \int_{0}^{\frac{1}{2}\pi} \sin^2\theta \cos^4\theta\, d\theta = 3a^2 \int_{0}^{\frac{1}{2}\pi} (\cos^4\theta - \cos^6\theta)\, d\theta$$

$$= 3a^2 \left\{ \frac{\pi}{2} \cdot \frac{1}{2} \cdot \frac{3}{4} - \frac{\pi}{2} \cdot \frac{1}{2} \cdot \frac{3}{4} \cdot \frac{5}{6} \right\}^{*} = \frac{3\pi a^2}{32}.$$

* v. *Elementary Calculus*, **6·6**.

Similarly, $\displaystyle\int_0^a xy\,dx = 3a^3\int_0^{\frac{1}{2}\pi}\sin^5\theta\cos^4\theta\,d\theta$

$$= -3a^3\int_0^{\frac{1}{2}\pi}(1-\cos^2\theta)^2\cos^4\theta\,d(\cos\theta)$$

$$= 3a^3(\tfrac{1}{5}-\tfrac{2}{7}+\tfrac{1}{9}) = \tfrac{8}{105}a^3.$$

Therefore $\bar{x} = \dfrac{256a}{315\pi}.$

We can calculate $\bar{y}$ in the same way, but since the equation of the curve is symmetrical in x and y, it is evident that $\bar{y} = \bar{x}$.

(ii) *Find the centre of gravity of a loop of the lemniscate* $r^2 = a^2\cos 2\theta$. The values of θ for which r vanishes are $\pm\tfrac{1}{4}\pi$.

Hence by using **10·32** (ii) we have

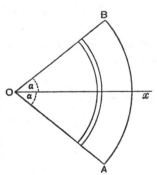

$$\bar{x} = \frac{\displaystyle\int_{-\frac{1}{4}\pi}^{\frac{1}{4}\pi}\tfrac{2}{3}r\cos\theta.\tfrac{1}{2}r^2\,d\theta}{\displaystyle\int_{-\frac{1}{4}\pi}^{\frac{1}{4}\pi}\tfrac{1}{2}r^2\,d\theta}.$$

Now the denominator

$$= \int_{-\frac{1}{4}\pi}^{\frac{1}{4}\pi}\tfrac{1}{2}a^2\cos 2\theta\,d\theta = \tfrac{1}{2}a^2,$$

and the numerator $= \displaystyle\int_{-\frac{1}{4}\pi}^{\frac{1}{4}\pi}\tfrac{1}{3}a^3\cos^3 2\theta\cos\theta\,d\theta,$

or, by putting $\sqrt{2}\sin\theta = \sin\phi$,   so that $\sqrt{2}\cos\theta\,d\theta = \cos\phi\,d\phi$,

the numerator $= \dfrac{1}{3\sqrt{2}}a^3\displaystyle\int_{-\frac{1}{2}\pi}^{\frac{1}{2}\pi}\cos^4\phi\,d\phi$

$$= \frac{2}{3\sqrt{2}}a^3.\frac{\pi}{2}.\frac{1}{2}.\frac{3}{4} = \frac{\pi a^3}{8\sqrt{2}},$$

and $\bar{x} = \dfrac{\pi a}{4\sqrt{2}}.$

It is evident by symmetry, or can be proved in like manner, that $\bar{y} = 0$.

(iii) *Find the centre of gravity of a sector of a circle in which the surface density varies as the distance from the centre.*

Let a be the radius and 2α the angle of the sector, AOB, and let $\rho = \lambda r$ denote the surface density, or mass per unit area, at distance r from the centre O.

Take the bisector of the angle AOB as axis of x. By symmetry the centre of gravity lies on this line.

In applying the formula

$$\bar{x} = \frac{\int x\rho\,dS}{\int\rho\,dS},$$

the choice of the element of area dS is limited by the consideration that the value of ρ must be the same at all points of the element when its area tends to vanish. But ρ has constant values along circular arcs of centre O; therefore we take for dS a narrow strip bounded by circles of radii r and $r+dr$, so that $dS = 2\alpha r\,dr$. Such a narrow strip may be regarded as a circular arc of radius r, so that the x co-ordinate of its centre of gravity is $\dfrac{r\sin\alpha}{\alpha}$ (10·311).

By substituting these values in the formula for $\bar{x}$, we get

$$\bar{x}=\frac{\displaystyle\int_0^a \frac{r\sin\alpha}{\alpha}.\lambda r\,.\,2\alpha r\,dr}{\displaystyle\int_0^a \lambda r\,.\,2\alpha r\,dr}=\frac{3}{4}\frac{a\sin\alpha}{\alpha}.$$

10·34. Students who are familiar with double and triple integrals will recognize at once that the formulae of **10·32** for plane surface distributions of matter can be expressed more generally by the forms

$$\bar{x}=\frac{\iint x\rho\,dx\,dy}{\iint \rho\,dx\,dy},\quad \bar{y}=\frac{\iint y\rho\,dx\,dy}{\iint \rho\,dx\,dy};$$

and in polar co-ordinates

$$\bar{x}=\frac{\iint r\cos\theta\,.\,\rho r\,d\theta\,dr}{\iint \rho r\,d\theta\,dr},\quad \bar{y}=\frac{\iint r\sin\theta\,.\,\rho r\,d\theta\,dr}{\iint \rho r\,d\theta\,dr}.$$

Also that there are similar formulae in triple integrals for finding the centres of gravity of volume distributions.

10·341. Volumes of Revolution.

To find the centre of gravity of a volume of revolution, i.e. a solid whose boundary is obtained by revolving a given curve $y=f(x)$ round the axis of x, cut off between plane ends $x=a$, $x=b$.

Since sections at right angles to the axis of x are circles, the solid may be regarded as built up of thin discs of area πy^2 and thickness dx, and the x-co-ordinate of the centre of gravity is given by

$$\bar{x}=\frac{\displaystyle\int_a^b x\,.\,\pi y^2\,dx}{\displaystyle\int_a^b \pi y^2\,dx},$$

and by symmetry the centre of gravity lies on the axis.

10·342. *Find the centre of gravity of a segment of height h of a sphere of radius a.*

Let O be the centre of the sphere and Ox an axis at right angles to the base of the segment meeting it in D and the curved surface in C, so that $DC = h$ and $OD = a - h$.

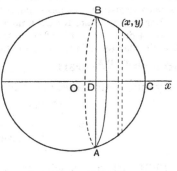

Taking slices of the segment parallel to its base and applying the formula of **10·341,**

$$\bar{x} = \frac{\int_{a-h}^{a} x \cdot \pi y^2 \, dx}{\int_{a-h}^{a} \pi y^2 \, dx}.$$

But x, y are co-ordinates of a point on a circle of radius a, viz. the section of the sphere by the plane of the paper; so that $y^2 = a^2 - x^2$, and

$$\bar{x} = \frac{\int_{a-h}^{a} (a^2 x - x^3) \, dx}{\int_{a-h}^{a} (a^2 - x^2) \, dx} = \frac{3}{4} \frac{(2a - h)^2}{3a - h}.$$

Cor. Putting $h = a$, we find that for a hemisphere, $\bar{x} = \tfrac{3}{8} a$.

10·343. Solid Octant of a Sphere. Consider an octant $OABC$ of radius a. Four such octants, as in the figure, make up a hemisphere whose centre of gravity G is on the radius OA at a distance $\tfrac{3}{8} a$ from its base. But the centres of gravity of the four octants are all at the same distance from the plane $BCB'C'$ and the plane containing them must pass through G the centre of gravity of the four. Hence and by symmetry the centre of gravity of a solid octant of the sphere is at a distance $\tfrac{3}{8} a$ from each of its plane faces.

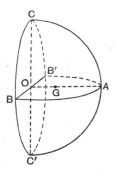

10·35. Centre of Gravity of a Zone of the Surface of a Sphere. Let $X'OX$ be a diameter of a circle and PM, QN ordinates to this diameter at neighbouring points P, Q on the circle, and let these ordinates meet a tangent $A'A$ parallel to $X'X$ in P', Q'.

Now let the figure revolve round $X'X$. The circle generates the surface of a sphere and the line $A'A$ traces out a cylinder which circumscribes the sphere. The ordinates MPP', NQQ' will trace out planes at right angles to $X'X$, between which planes lie a slice of the sphere and the corresponding slice of the cylinder.

The area of the narrow zone of the sphere traced out by the arc PQ is $2\pi MP.PQ$, and the area of the corresponding zone on the circumscribing cylinder is

$$2\pi MP'.P'Q'.$$

But

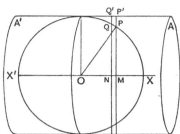

$$\frac{P'Q'}{PQ}=\sin QPP'=\sin POM$$

$$=\frac{MP}{OP}=\frac{MP}{MP'},$$

therefore

$$MP'.P'Q'=MP.PQ.$$

Therefore the areas of corresponding narrow zones of the sphere and the circumscribing cylinder are equal, and by addition the same must be true for zones of any width. Therefore the area of the surface of a sphere intercepted between parallel planes is equal to the circumference of the sphere multiplied by the distance between the planes.

Now to find the centre of gravity of such a zone of a sphere; let it be divided into narrow zones by a large number of parallel planes at equal distances apart, then the areas intercepted between consecutive planes are all equal and their centres of gravity are uniformly distributed along the axis of figure, so the problem is that of finding the centre of gravity of a uniform distribution of equal particles in the same straight line, and it is midway between the end particles, i.e at the point on the axis of the zone midway between its bounding planes.

Cor. i. The centre of gravity of the surface of a segment of a sphere is at a distance from the centre of its base equal to half its height.

Cor. ii. The centre of gravity of a thin hemispherical shell is at the middle point of its central radius.

Cor. iii. The centre of gravity of the curved surface of an octant of a sphere is at a distance equal to half the radius from each of the three perpendicular planes which bound the octant.

10·36. Areas on a Sphere. Let dS be a small element of area on the surface of a sphere of radius a, $d\Pi$ its projection on a plane through the centre O and z its distance from this plane.

Then $d\Pi$ is equal to dS multiplied by the cosine of the angle between $d\Pi$ and the tangent plane at a point of dS, i.e.

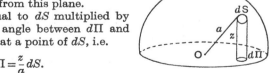

$$d\Pi=\frac{z}{a}dS.$$

Now if dS is a part of a finite area S on the surface of the sphere, we have for the centre of gravity of S

$$\bar{z} = \frac{\int z \, dS}{\int dS} = \frac{\int a \, d\Pi}{\int dS}$$

or

$$\frac{\bar{z}}{a} = \frac{\Pi}{S},$$

where Π is the projection of S.

Cor. We obtain the result of **10·35**, Cor. (ii) at once by taking S to be the surface of a hemisphere of area $2\pi a^2$, its projection Π being a circle of radius πa^2.

10·4. Applications of Orthogonal Projection. Among simple properties of figures obtained by orthogonal projection are the following:

(i) parallel lines project into parallel lines;

(ii) the ratio of two segments of the same line or of parallel lines is unaltered by orthogonal projection;

(iii) the centre of gravity of a plane area projects orthogonally into the centre of gravity of the projection.

The truth of the first two statements is evident from the figure in which FAB and KCD are parallel lines on an inclined plane $FEML$

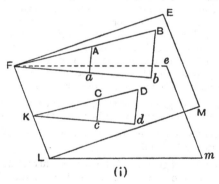

(i)

and Fab, Kcd are their projections on a horizontal plane $FemL$. The projections Fb and Kd are parallel, for if they had a common point it would be the projection of a common point of FB and KD. Also since FB and KD are parallel they are equally inclined to the parallel lines Fb and Kd, and then by similar figures

$$ab : cd = AB : CD.$$

To prove (iii), let S be any area on a plane inclined at an angle α to a horizontal plane, and Π the projection of S on the horizontal plane, fig. (ii).

Let S be divided into small elements by drawing lines of greatest slope and lines parallel to the intersections of the planes. It follows that, if $d\Pi$ is the projection of an element dS, then $d\Pi = \cos\alpha \cdot dS$.

Also if $\bar{x}, \bar{y}, \bar{z}$ are the co-ordinates of the centre of gravity of S referred to the horizontal axes Ox, Oy and a vertical axis as shewn in the figure, and x, y, z are co-ordinates of a point of dS, then

$$\bar{x} = \frac{\int x\,dS}{\int dS} = \frac{\int x \sec \alpha\,d\Pi}{\int \sec \alpha\,d\Pi}$$

$$= \frac{\int x\,d\Pi}{\int d\Pi}, \quad \text{and similarly} \quad \bar{y} = \frac{\int y\,d\Pi}{\int d\Pi}.$$

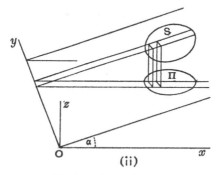

(ii)

But x, y, 0 are co-ordinates of a point of $d\Pi$, so that the last two fractions are also the x, y co-ordinates of the centre of gravity of Π and so the centre of gravity of S projects into the centre of gravity of Π.

10·41. Quadrant of an Ellipse. A quadrant of a circle can be projected orthogonally into a quadrant of an ellipse. Thus if OAB, fig. (i), be a quadrant of a circle and it be projected on to a plane through OA, all lines parallel to OA are unaltered in length, and all lines parallel to OB are reduced in the same ratio.

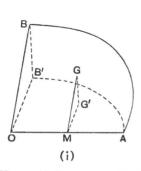

(i)

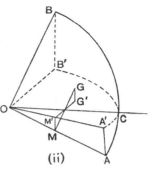

(ii)

Hence, if GM is an ordinate to OA from the centre of gravity, and B' and G' are the projections of B and G, we have

$$\frac{MG'}{OB'} = \frac{MG}{OB} = \frac{4}{3\pi}, \quad \textbf{10·322.}$$

Therefore if $OA = a$, $OB' = b$ are the semiaxes of the elliptic quadrant OAB' and $\bar{x}$, $\bar{y}$ the co-ordinates of its centre of gravity G' referred to OA, OB' as axes, we have

$$\frac{\bar{x}}{a} = \frac{\bar{y}}{b} = \frac{4}{3\pi}.$$

Again, in fig. (ii), by projecting the quadrant OAB of a circle on to a plane which cuts the plane of the quadrant in any radius OC, we get a quadrant $OA'B'$ of an ellipse bounded by conjugate radii OA', OB', and the ordinate MG which is parallel to OB projects into a line $M'G'$ parallel to OB', and

$$\frac{OM'}{OA'} = \frac{OM}{OA} = \frac{4}{3\pi},$$

by similar triangles and **10·322**; while

$$\frac{M'G'}{OB'} = \frac{MG}{OB} = \frac{4}{3\pi}$$

(**10·4** (ii) and **10·322**).

Hence if $OA' = a'$ and $OB' = b'$ are the lengths of conjugate radii and $\bar{x}'$, $\bar{y}'$ are oblique co-ordinates of G' parallel to the conjugate radii

$$\frac{\bar{x}'}{a'} = \frac{\bar{y}'}{b'} = \frac{4}{3\pi}.$$

10·5. Theorems of Pappus.

(i) *If a plane curve rotates through any angle about an axis in its plane which does not cross it, the area of the surface traced out by the curve is equal to the length of the curve multiplied by the length of the path of its centre of gravity.*

For if ds denotes an element of arc of the curve at a distance y from the axis of rotation and θ the angle through which it turns, the element ds traces out a narrow ribbon of length $y\theta$ and area $y\theta\,ds$. Therefore the area of the surface traced out by the whole curve is represented by

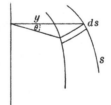

$$\int y\theta\,ds = \theta \int y\,ds$$
$$= \theta\bar{y} \times s,$$

where s is the length of the curve and $\bar{y}$ is the distance of its centre of gravity from the axis, and this proves the theorem.

(ii) *If a plane area rotates through any angle about an axis in its plane which does not cross it, the volume traced out is equal to the area multiplied by the length of the path of its centre of gravity.*

For if dS denotes an element of the area at a distance y from the axis of rotation and θ the angle through which it turns, the element dS traces out a narrow tube of length $y\theta$, cross section dS and volume $y\theta\,dS$. Therefore the volume traced out by the whole surface is represented by

$$\int y\theta\,dS = \theta \int y\,dS$$
$$= \theta\bar{y} \times S,$$

where S is the given area and $\bar{y}$ is the distance of its centre of gravity from the axis.

10·51. Examples. (i) *To find the area of the surface and the volume of an anchor ring obtained by rotating a circle of radius a about an axis in its plane at a distance c from its centre, where $c > a$.*

From **10·5** (i) the area of the surface of the anchor ring

$$= 2\pi a \times 2\pi c = 4\pi^2 ac;$$

and from **10·5** (ii) the volume

$$= \pi a^2 \times 2\pi c = 2\pi^2 a^2 c.$$

(ii) *Use the theorems of Pappus to find the centre of gravity of*

(a) *a semicircular arc,*

(b) *a semicircular area.*

In each case let $\bar{x}$ denote the distance of the centre of gravity from the centre, and a the radius of the given circle, and let a semicircular arc and area make a complete revolution about its diameter.

In both cases the figure traced out in a complete revolution is a sphere. Therefore

(a) $\qquad 2\pi\bar{x} \times \pi a = \text{area of sphere} = 4\pi a^2,$

so that $\qquad\qquad\qquad \bar{x} = \dfrac{2a}{\pi};$

and (b) $\qquad 2\pi\bar{x} \times \tfrac{1}{2}\pi a^2 = \text{volume of sphere} = \tfrac{4}{3}\pi a^3,$

so that $\qquad\qquad\qquad \bar{x} = \dfrac{4a}{3\pi}.$

10·6. Lagrange's Formula.

If G be the centre of gravity of a system of particles of masses $m_1, m_2, m_3, \ldots$ situated at the points $A_1, A_2, A_3, \ldots$ and O be any other point, then

$$\Sigma m\,OA^2 = \Sigma m\,GA^2 + OG^2 \Sigma m.$$

For any point A, let ON be the projection of OA on the line OG.

Then

$$\Sigma m\, OA^2 = \Sigma m\,(GA^2 + OG^2 + 2OG\,.\,GN)$$
$$= \Sigma m\, GA^2 + OG^2\Sigma m$$
$$+ 2OG\Sigma m\, GN.$$

But if we take G as an origin and OG produced as an axis of x, so that GN is the x of the point A, then

$$\Sigma m\, GN = \Sigma m x = \bar{x}\Sigma m = 0,$$

since the centre of gravity is the origin: therefore

$$\Sigma m\, OA^2 = \Sigma m\, GA^2 + OG^2\Sigma m.$$

This theorem is only a slightly modified form of the theorem of parallel axes in the theory of moments of inertia*; it is useful in metrical geometry.

10·61. Example. *A uniform triangular lamina ABC of weight W is suspended from a fixed point by strings of lengths l_1, l_2, l_3 attached to its corners A, B, C. Shew that the tensions in the strings are*

$$Wl_1/\{3(l_1{}^2 + l_2{}^2 + l_3{}^2) - a^2 - b^2 - c^2\}^{\frac{1}{2}}$$

and similar expressions.

Let T_1, T_2, T_3 denote the tensions in the strings OA, OB, OC. Since their resultant is W acting in the vertical OG, where G is the centre of gravity of the lamina, and G is also the centre of gravity of particles of equal mass at A, B, C, therefore, by **3·43**,

$$\frac{T_1}{l_1} = \frac{T_2}{l_2} = \frac{T_3}{l_3} = \frac{W}{3OG} \quad \ldots\ldots(1).$$

But by **10·6** $l_1{}^2 + l_2{}^2 + l_3{}^2 = AG^2 + BG^2 + CG^2 + 3OG^2,$

and, by completing a parallelogram of which BG, GC are adjacent sides and observing that its diagonals are of length a and AG,

$$AG^2 + a^2 = 2(BG^2 + CG^2),$$

similarly $BG^2 + b^2 = 2(CG^2 + AG^2),$

and $CG^2 + c^2 = 2(AG^2 + BG^2),$

so that $AG^2 + BG^2 + CG^2 = \tfrac{1}{3}(a^2 + b^2 + c^2).$

Therefore $9OG^2 = 3(l_1{}^2 + l_2{}^2 + l_3{}^2) - a^2 - b^2 - c^2,$

and the required result follows from (1).

* v. *Dynamics*, **13·2**.

10·7. Miscellaneous Examples.

(i) *The co-ordinates of the extremities of the bounding radii of a sector of the ellipse $x^2/a^2 + y^2/b^2 = 1$ are x_1, y_1, and x_2, y_2. Prove that the co-ordinates of its centre of gravity are*

$$\frac{2a}{3b}\cdot\frac{y_2-y_1}{\phi}, \quad and \quad \frac{2b}{3a}\cdot\frac{x_1-x_2}{\phi},$$

where $\qquad ab\sin\phi = x_1 y_2 - x_2 y_1.$ [S.]

Let OPQ be the sector of the ellipse, OA, OB its semiaxes of lengths a, b, and P', Q', B' the points on the auxiliary circle corresponding to P, Q, B.

Then the elliptic figure may be regarded as an orthogonal projection of the circular figure in which every ordinate has been reduced in the ratio $a:b$; and if G' is the centre of gravity of the circular sector $OP'Q'$ and G that of the elliptic sector OPQ, G lies on the ordinate $G'L$ and is such that

$$GL:G'L = b:a.$$

Now if α and $\alpha + \phi$ are the eccentric angles of the points P and Q, i.e. the angles $P'OA$, $Q'OA$, the angle $P'OQ' = \phi$ and

$$OG' = \tfrac{4}{3}a\,\frac{\sin\tfrac{1}{2}\phi}{\phi} \quad (\mathbf{10\cdot321}).$$

Also $\qquad x_1 y_2 - x_2 y_1 = ab\{\cos\alpha\sin(\alpha+\phi) - \cos(\alpha+\phi)\sin\alpha\}$

$$= ab\sin\phi;$$

identifying $P'OQ'$ with the ϕ of the question.

Then the co-ordinates of G are given by

$$\bar{x} = OL = OG'\cos(\alpha + \tfrac{1}{2}\phi)$$

$$= \frac{4}{3}a\,\frac{\sin\tfrac{1}{2}\phi\cos(\alpha+\tfrac{1}{2}\phi)}{\phi}$$

$$= \frac{2}{3}a\,\frac{\sin(\alpha+\phi)-\sin\alpha}{\phi}$$

$$= \frac{2}{3}\frac{a}{b}\,\frac{y_2-y_1}{\phi};$$

and $\qquad \bar{y} = GL = \dfrac{b}{a}G'L = \dfrac{b}{a}\,OG'\sin(\alpha+\tfrac{1}{2}\phi)$

$$= \frac{4}{3}b\,\frac{\sin\tfrac{1}{2}\phi\sin(\alpha+\tfrac{1}{2}\phi)}{\phi}$$

$$= \frac{2}{3}b\,\frac{\cos\alpha-\cos(\alpha+\phi)}{\phi}$$

$$= \frac{2}{3}\frac{b}{a}\,\frac{x_1-x_2}{\phi}.$$

(ii) *An isosceles triangular lamina is such that its mass per unit area at every point is proportional to the product of the distances of the point from the equal sides of the triangle. Prove that the distance of the centre of gravity from the vertex is four-fifths of the altitude.*

Let OAB be the triangle in which $OA = OB$, let $OC = a$ be the altitude and let the angle AOB be 2α.

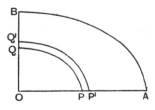

Take the axis of x along OC. Let P be the point (x, y) and PM, PN perpendicular to OA, OB. Then it is easy to see that

$$PM = x \sin \alpha - y \cos \alpha$$

and $$PN = x \sin \alpha + y \cos \alpha,$$

so that the mass per unit area is proportional to $\quad x^2 \sin^2 \alpha - y^2 \cos^2 \alpha.$

By symmetry the centre of gravity lies on Ox, and its distance from O is given by

$$\bar{x} = \frac{\displaystyle\int_0^a \int_{-x \tan \alpha}^{x \tan \alpha} x(x^2 \sin^2 \alpha - y^2 \cos^2 \alpha)\, dx\, dy}{\displaystyle\int_0^a \int_{-x \tan \alpha}^{x \tan \alpha} (x^2 \sin^2 \alpha - y^2 \cos^2 \alpha)\, dx\, dy}$$

$$= \frac{\displaystyle\int_0^a (2x^4 \tan \alpha \sin^2 \alpha - \tfrac{2}{3} x^4 \tan^3 \alpha \cos^2 \alpha)\, dx}{\displaystyle\int_0^a (2x^3 \tan \alpha \sin^2 \alpha - \tfrac{2}{3} x^3 \tan^3 \alpha \cos^2 \alpha)\, dx}$$

$$= \frac{\displaystyle\int_0^a x^4\, dx}{\displaystyle\int_0^a x^3\, dx} = \frac{4}{5} a.$$

(iii) *The mass per unit area at any point of a lamina bounded by a quadrant of an ellipse and its axes is proportional to the nth power of the linear dimensions of the similar and similarly situated ellipse which passes through the point. Find the centre of gravity.*

Let OAB be the lamina, and let

$$\frac{x^2}{a^2} + \frac{y^2}{k^2 a^2} = 1$$

be the equation of the boundary ellipse AB.

The equation of a similar ellipse PQ is then

$$\frac{x^2}{a'^2} + \frac{y^2}{k^2 a'^2} = 1.$$

The area of the quadrant OPQ is $\tfrac{1}{4} \pi k a'^2$, and by **10·41** the x co-ordinate of the centre of gravity of this quadrantal area is $\dfrac{4a'}{3\pi}$, so that the x-moment of this quadrantal area is $\frac{1}{3}ka'^3$.

Now consider a narrow strip of the lamina bounded by the elliptic quadrant PQ of semiaxes a', ka' and a neighbouring quadrant $P'Q'$ of semiaxes $a' + da'$, $k(a' + da')$. The area of this strip is, to the first order, the differential of the area OPQ, i.e. $\frac{1}{2}\pi ka'\,da'$, but the mass per unit area is constant over this narrow strip and proportional to a'^n, say $\lambda a'^n$, so that the mass of the strip $PQQ'P'$ is $\frac{1}{2}\pi k\lambda a'^{n+1}\,da'$, and the mass of the lamina OAB

$$= \tfrac{1}{2}\pi k\lambda \int_0^a a'^{n+1}\,da' = \frac{1}{2(n+2)}\pi k\lambda a^{n+2}.$$

Again the x-moment of the area of strip $PQQ'P'$ is, to the first order, the differential of the x-moment of the area OPQ, i.e. $ka'^2\,da'$, therefore the x-moment of the mass of the strip $PQQ'P'$ is $k\lambda a'^{n+2}\,da'$, and the x-moment of the mass of the lamina OAB

$$= k\lambda \int_0^a a'^{n+2}\,da' = \frac{1}{n+3}k\lambda a^{n+3}.$$

Hence, by dividing the x-moment by the mass, we find that

$$\bar{x} = \frac{2(n+2)}{\pi(n+3)}a;$$

and similarly

$$\bar{y} = \frac{2(n+2)}{\pi(n+3)}ka.$$

EXAMPLES

1. From a circular area of radius a a smaller circular area one quarter the size is so stamped out that its centre bisects a radius of the larger circle. Find the centre of gravity of the remainder.   [S.]

2. If the C.G. of a quadrilateral is the same as that of four equal particles placed at its angular points, shew that the quadrilateral must be a parallelogram.   [C.]

3. Three rods of unequal length are joined together to form a triangle ABC. If the masses are equal, prove that the centre of gravity coincides with that of the area. If the masses of the sides a, b, c are proportional to $b + c - a$, $c + a - b$, $a + b - c$, prove that the centre of gravity is the centre of the inscribed circle.   [I.]

4. $ABCD$ is a heavy uniform square plate, the portion CBH is removed, where H is a point in AB. The remainder is placed with its plane vertical and AH in contact with a smooth horizontal plane; shew that equilibrium will be impossible unless AH/AB is greater than $(\sqrt{3} - 1)/2$.   [S.]

5. A triangle of uniform rods of different densities has its centre of

gravity at the centre of its circumscribing circle. Prove that the densities are proportional to the secants of the opposite angles.   [S.]

6. Prove that the centre of gravity of the perimeter of a triangle is the centre of the inscribed circle of the triangle formed by joining the middle points of the sides.

If the sides are of different densities and the centre of gravity of the perimeter is the nine-points centre of the triangle, then the densities of the sides are proportional to the cosines of the opposite angles. [S.]

7. Particles, whose masses are proportional to $\sin A \cos (B - C)$, $\sin B \cos (C - A)$, $\sin C \cos (A - B)$, are placed at the vertices of a triangle ABC. Prove that their centre of gravity lies at the centre of the nine-points circle of the triangle.   [S.]

8. Three wires of uniform but different densities form a triangle ABC. If the centre of gravity is at the centre of the inscribed circle, shew that the masses of the wires are in the ratios

$$\cot \frac{A}{2} : \cot \frac{B}{2} : \cot \frac{C}{2}.$$ [S.]

9. A uniform solid consists of a cone and a hemisphere fastened together so that their plane faces coincide, the diameter of the hemisphere being equal to that of the base of the cone. Shew that if the height of the cone does not exceed $\sqrt{3}$ times the radius of the base, the solid will always move to an upright position if placed with the surface of the hemisphere on a horizontal plane.   [C.]

10. Find the centre of mass of a rod whose density varies as the square of the distance from one end.   [S.]

11. Prove that the centroid of the surface of a tetrahedron is the centre of the sphere inscribed in the tetrahedron formed by joining the centroids of the faces.   [S.]

12. A body is bounded by two spheres (radii a, b) which touch internally. Find the centre of gravity and shew that in the limiting case when $a = b$ it divides the common diameter in the ratio of $2 : 1$.   [S.]

13. A piece of wire of given length is bent into the form of a circular quadrant and its two bounding radii; find the centre of gravity of the whole.   [C.]

14. Two tangents are drawn to a circle of radius a subtending an angle 2α at the centre. Prove that the centre of gravity of the figure bounded by the tangents and the smaller arc between them is at a distance from the centre equal to

$$\frac{a}{3} \cdot \frac{\tan^2 \alpha \sin \alpha}{\tan \alpha - \alpha}.$$ [S.]

15. $ABCD$ is a quadrilateral, AC and BD intersect in E, points F and G are taken on AC and BC, such that AF and BG are equal to EC and ED respectively; prove that the centres of gravity of the triangles FBD and GAC coincide.   [I.]

16. $ABCD$ is a uniform plane quadrilateral lamina, whose diagonals intersect in E. If the point H divides AC in the ratio

$$AC + EC : AC + EA,$$

and K divides BD in the ratio

$$BD + ED : BD + EB,$$

shew that the centre of gravity of the lamina bisects HK. [S.]

17. Weights P, Q, R at the angular points of a triangle ABC have their centre of gravity at the centre of the inscribed circle. Prove that, ρ and r being the radii of the circumscribed and inscribed circles,

$$2\rho r (P + Q + R)^2 = QR . a^2 + RP . b^2 + PQ . c^2.$$ [S.]

18. A wire in the form of a circular arc AB of line density w, with a length l of string of the same line density hanging vertically from A, rests in equilibrium in contact with a fixed smooth horizontal peg. Prove that l is equal to AT, where T is the point in which the horizontal plane through B meets the tangent to the arc at A. [I.]

19. A frustum is cut from a right circular cone of semivertical angle α by planes at right angles to the axis and at distance h and h' from the vertex. Prove that the distance from the vertex of the centre of gravity of the whole surface of the frustum is

$$\frac{1}{3} \cdot \frac{h^3 (2 \sec \alpha + 3 \tan \alpha) - h'^3 (2 \sec \alpha - 3 \tan \alpha)}{h^2 (\sec \alpha + \tan \alpha) - h'^2 (\sec \alpha - \tan \alpha)}.$$ [S.]

20. Two pieces of the same uniform wire, one a semi-circle and the other an arc of a circle subtending an angle $2\theta \, (< \pi)$ at its centre, are placed so as to form a crescent. If θ is such that the centre of gravity is on the inner arc, shew that it is given by the equation

$$\sin^2 \theta + \sin \theta = \theta + \frac{\pi}{2} \sin \theta (1 - \cos \theta).$$ [S.]

21. Two parallel chords of a circle of radius a subtend angles 2α, 2β at the centre. Shew that the centre of gravity of the part of the disc between the chords is at a distance $\dfrac{2a^3 (\sin^3 \alpha - \sin^3 \beta)}{3A}$ from the centre, where A is the area of the figure. [S.]

22. A solid homogeneous hemisphere of radius a and weight W rests with its curved surface on a fixed horizontal plane, and a particle of weight W' is placed on it at a distance x from the centre; prove that, in the position of equilibrium, the friction exerted between the particle and the hemisphere is equal to

$$8x W'^2 (9a^2 W^2 + 64x^2 W'^2)^{-\frac{1}{2}}.$$ [S.]

23. A thin hemispherical bowl of weight W contains a weight W' of water and rests on a rough inclined plane of inclination α. Shew that the plane of the top of the bowl makes an angle ϕ with the horizontal given by

$$W \sin \phi = 2 (W + W') \sin \alpha.$$ [S.]

24. If a line is drawn cutting two sides AB and AC of a given uniform triangular plate in P and Q so that the area $BPQC$ is constant, prove that the locus of the centre of mass of this area is an arc of a hyperbola.   [S.]

25. A frustum of a uniform right circular cone whose semivertical angle is α is made by cutting off $1/n$th of the axis; prove that the frustum will rest with a slant side on a horizontal plane if

$$\tan^2\alpha < \frac{1}{4}\frac{3n^4-4n^3+1}{n^3-1}.$$ [S.]

26. A sphere of radius a is inscribed in a cone of vertical angle 2α. Shew that the distance between the centre of the sphere and the centre of gravity of that part of the volume of the cone which lies between its vertex and the nearer part of the surface of the sphere is

$$\frac{a(1+\sin\alpha)^2}{4\sin\alpha}.$$ [S.]

27. Find the co-ordinates (referred to the axes of the curve) of the centre of gravity of the smaller portion of an elliptic lamina cut off by a chord joining an end of one axis to an end of the other.   [S.]

28. A frustum of a cone (vertical angle $120°$) is cut off by two spheres whose centres are at the vertex of the cone, the radius of one being twice that of the other. If the density vary as the distance from the vertex, shew that the centre of mass of the frustum divides its axis in the ratio $6:19$.   [S.]

29. Having given two triangles ABC, $A'B'C'$ in the same plane, prove that three masses can be found whose ratios have unique values, such that if they are placed at the points A, B, C or at the points A', B', C' their centroid is at the same point in either case; one of the masses may be negative.   [S.]

30. A portion of a uniform thin spherical shell is bounded by a plane circular rim. When the shell is freely suspended from a point in the rim, the radius through the lowest point of the rim is horizontal; shew that the plane bounding the shell is at a distance from the centre of the sphere equal to two-thirds of the radius.   [S.]

31. $ABCD$ is a tetrahedron: a point B' is taken in AB so that $AB'=\frac{4.5}{7.6}AB$, and points C', D' are taken in a similar manner in AC, AD. Shew that, if the tetrahedron be divided into two parts by a plane parallel to BCD and equidistant from it and the point A, then the centre of mass of the larger part coincides with that of equal masses at B', C', D'.   [S.]

32. The lengths of the parallel edges of a homogeneous prismatic solid are represented by the numbers 2, 1, 1 respectively. Prove that the distances of the centre of gravity of the solid and of the longer edge from the face containing the equal edges are in the ratio $3:8$.   [S.]

33. A solid sector of semiangle α is cut from a solid sphere and replaced with its edge in the same position as before but with its vertex outwards; shew that the centre of gravity of the resulting system will coincide with the centre of the sphere if $\cos\alpha = \frac{3}{5}$. [S.]

34. If G is the centre of gravity of a solid cone on an elliptic base and P is any point in the circumference of the base, then the generating line of the cone cut by PG produced is divided in the ratio of 3 to 2. [S.]

35. A leaning tower of n equal coins is constructed on a horizontal table, so that the centres of gravity of all the coins lie in one straight line; find the greatest inclination of this line to the vertical. [C.]

36. The thickness of a thin circular homogeneous plate at any point is proportional to the distance of the point from a tangent to its perimeter. Find the volume of the plate and the position of its c.g., taking a to be its radius and t to be its thickness at the centre. [S.]

37. If the density of any point of an arc of a uniform circular wire varies as its distance from the central radius, prove that the centre of mass is at a point on the central radius, midway between the chord and the arc. [S.]

38. An isosceles triangular lamina is such that its mass per unit area at every point is proportional to the sum of the distances of the point from the equal sides of the triangle. Prove that the distance of the centre of gravity from the vertex is three-fourths of the altitude.

39. Find the centre of gravity of semicircular plate of radius a whose mass per unit area at any point varies as $\sqrt{(a^2 - r^2)}$, where r is the distance of the point from the centre. [S.]

40. The density at any point in a sector of a circle varies as the distance from the centre. Find the centre of gravity of the sector. [S.]

41. The mass per unit length at any point of a straight beam of length l is $\rho\,(1 + x/l)$, where x is the distance of the point from one end of the beam. Find (i) the mass of the beam, (ii) the position of its centre of mass.

If the beam rests horizontally on supports at its ends, prove that the bending moment is greatest at a distance $(\frac{1}{3}\sqrt{21} - 1)\,l$ from one end. [I.]

42. Find the centre of gravity of a plate in the form of a quadrant AOB of an ellipse, the thickness at any point of the plate varying as the product of the distances of the point from OA and OB. [I.]

43. An elliptic section is made of a solid circular cone. If G be the centre of gravity of the part between this section and the vertex V, and if GM be perpendicular to the axis, prove that

$$VM = \tfrac{3}{8}(a + a')\cos\alpha, \quad \text{and} \quad MG = \tfrac{3}{8}(a - a')\sin\alpha,$$

where 2α is the vertical angle of the cone, and a and a' are the longest and shortest generators.

If H be the centre of gravity of the curved surface, and HN be perpendicular to the axis, prove also that

$$VN = \frac{\cos\alpha}{6\,(a+a')}\,(3a^2 + 2aa' + 3a'^2),$$

$$NH = \tfrac{1}{2}\,(a-a')\sin\alpha. \qquad\qquad [\text{I.}]$$

44. A light equilateral triangular frame is loaded at the corners with weights w_1, w_2, w_3 and suspended from a fixed point by strings of lengths l_1, l_2, l_3 attached to its corners. Prove that the tensions T_1, T_2, T_3 in the strings are given by

$$\frac{w_1 l_1}{T_1} = \frac{w_2 l_2}{T_2} = \frac{w_3 l_3}{T_3}$$

$$= \frac{\{(w_1+w_2+w_3)(w_1 l_1{}^2 + w_2 l_2{}^2 + w_3 l_3{}^2) - (w_2 w_3 + w_3 w_1 + w_1 w_2)\,a^2\}^{\frac{1}{2}}}{w_1 + w_2 + w_3},$$

where a is the length of a side of the triangle. [S.]

ANSWERS

1. $\tfrac{1}{6}a$ from centre.

10. $\tfrac{3}{4}$ length of rod from given end.

12. $(a^3 + a^2 b + ab^2 + b^3)/(a^2 + ab + b^2)$ from the point of contact.

13. $4\sqrt{2}a/(\pi+8)$ from the centre.

27. $2a/3\,(\pi-2)$, $2b/3\,(\pi-2)$.

35. $\tan^{-1}\{a/(n-1)\,b\}$, where a is the radius and $2b$ the thickness of a coin.

36. $\pi a^2 t$; $\tfrac{1}{4}a$ from centre.

39. $\tfrac{3}{8}a$ from centre.

40. $\dfrac{3}{4}\dfrac{a\sin\alpha}{\alpha}$ from centre.

41. $\tfrac{3}{2}\rho l$; $\bar{x} = \tfrac{5}{9}l$.

42. $\bar{x} = \tfrac{8}{15}OA$, $\bar{y} = \tfrac{8}{15}OB$.

Chapter XI

WORK AND ENERGY

11·1. The **work done by a force** when its point of application undergoes a small displacement is defined to be the product of the force and the orthogonal projection of the displacement upon the line of action of the force.

Thus if the point of application A of a force P receives a small displacement to A' where AA' makes an angle θ with the direction of the force P, and AN is the orthogonal projection of AA' upon the line of action of P, the work done by the force P in the displacement AA' is defined to be $P.AN$ or $P.AA'\cos\theta$, and is positive, zero or negative according as θ is an acute angle, a right angle or an obtuse angle.

11·11. Scalar Products. If A, B are any two vectors inclined to one another at an angle θ, then $AB\cos\theta$ is defined to be *the scalar product* of the vectors A, B and is variously denoted by (AB), or $A.B$ or simply AB.

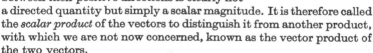

This product of the magnitudes of the vectors multiplied by the cosine of the angle between their positive directions is clearly not a directed quantity but simply a scalar magnitude. It is therefore called the *scalar product* of the vectors to distinguish it from another product, with which we are not now concerned, known as the vector product of the two vectors.

We observe that the work done by a force in a small displacement is the scalar product of the force and the displacement, and that work is *not* a vector.

11·12. *When any number of forces act upon a particle which undergoes a small displacement the algebraical sum of the work* *done by the separate forces is equal to the work done by their resultant.*

Let O be the particle which is displaced to O'. Let P be any one of the forces and R the resultant force, and let θ, ϕ be the angles which they make with OO'.

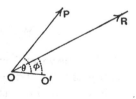

Then the algebraical sum of the work done by the forces of which P is the type $= \Sigma P . OO' \cos \theta$

$$= OO' . \Sigma P \cos \theta$$
$$= OO' . R \cos \phi$$
$$= R . OO' \cos \phi$$
$$= \text{work done by the resultant force.}$$

11·13. If in **11·12** the forces are P_1, P_2, P_3, ..., and dp_1, $dp_2, dp_3, ...$ denote the projections of OO' on the lines of action of the forces, and dr denotes the projection of OO' on the line of action of R, then what we have proved is that

$$\Sigma P \, dp = R \, dr.$$

Thus, in particular, if X, Y, Z denote the rectangular components of a resultant force R acting on a particle which receives any small displacement whose projections on the coordinate axes and on R are dx, dy, dz, and dr, we have

$$X \, dx + Y \, dy + Z \, dz = R \, dr.$$

11·14. Displacement along a Curve. Conservative Field of Force. Now consider the displacement of a particle along a continuous curve AB under the action of a given field of force; meaning by the latter term that at whatever point the particle may be situated there is a definite force acting upon it known in magnitude and direction. Let R denote this force when the particle is at

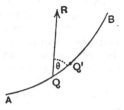

a point Q on the curve; let θ be the angle at Q between R and the curve and let QQ' be an infinitesimal element ds of the curve. Then $R \cos \theta \, ds$ denotes the work done in the displacement QQ', and the work done as the particle moves along the curve from A to B may be denoted by $\int_A^B R\cos\theta\,ds$ or by $\int_A^B R\,dr$, where dr denotes the projection of ds upon the line of action of R, and this by **11·13** is equivalent to

$$\int_A^B (X\,dx + Y\,dy + Z\,dz).$$

If the integrand $X\,dx + Y\,dy + Z\,dz$ is an exact differential dW of some single-valued function of position W, then the integral is $[W]^B_A$, i.e. the excess of the value of W at B over its value at A; and the value of the integral does not depend upon the particular path by which the particle has been moved from A to B but is the same for all paths. In such a case W is called the *work function* and the field of force is said to be a *conservative field*. To emphasize this definition we repeat that *a field of force is conservative if the work done by the forces of the field in the transit of a particle from one point of the field to another depends only on the positions of these points and not upon the particular path of transit.*

From **11·13** it appears that the differential of the work function can be expressed in the form $\Sigma P\,dp$, i.e. the sum of each force multiplied by the projection of the displacement of its point of application upon its line of action, and that when this expression $\Sigma P\,dp$ is an exact differential of a single-valued function then the forces constitute a conservative system.

11·15. Examples. (i) *The Earth's gravitational field is conservative.*

Let a particle move along a curve AB in a vertical plane under the action of no force but its weight w. Taking horizontal and vertical axes Ox, Oy in the plane of the curve, the force components in any position are $X = 0$, $Y = -w$; therefore the work done as the particle moves from A to B

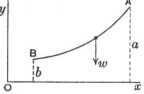

$$= \int (X\,dx + Y\,dy)$$
$$= \int_a^b -w\,dy$$
$$= w(a-b),$$

where a, b are the ordinates of the points A, B.

Hence the work done is independent of the actual path and the field of force is conservative.

(ii) *A field of force in which the forces at any point are directed towards fixed points and are proportional to some powers of the distances from these points is conservative.*

Let P be any point in the field and $r_1, r_2, r_3, \ldots$ its distances from the fixed points $O_1, O_2, O_3, \ldots$.

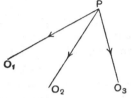

Let the forces acting along PO_1, PO_2, $PO_3, \ldots$ be $\lambda_1 r_1{}^{n_1}, \lambda_2 r_2{}^{n_2}, \lambda_3 r_3{}^{n_3}, \ldots$; then the work done by these forces in a small displacement of P whose projections on O_1P, O_2P, $O_3P, \ldots$ are $dr_1, dr_2, dr_3, \ldots$ is

$$-\lambda_1 r_1{}^{n_1} dr_1 - \lambda_2 r_2{}^{n_2} dr_2 - \lambda_3 r_3{}^{n_3} dr_3 - \ldots,$$

which is the differential of the function

$$-\frac{\lambda_1}{n_1+1} r_1{}^{n_1+1} - \frac{\lambda_2}{n_2+1} r_2{}^{n_2+1} - \frac{\lambda_3}{n_3+1} r_3{}^{n_3+1} - \ldots.$$

Therefore the field of force is conservative.

11·16. When a particle moves round a closed path in a conservative field of force, the work function obtained by integrating round the path returns to its initial value on completing the circuit so that the total work done is zero. This would not necessarily be the case if the function were not single-valued. For example, if we had

$$X\,dx + Y\,dy + Z\,dz = \frac{x\,dy - y\,dx}{x^2 + y^2}$$

$$= d\left(\tan^{-1}\frac{y}{x}\right);$$

this is an exact differential, but of a many-valued function; it is, in fact, in plane polar co-ordinates $d\theta$, and for any closed path which includes the origin it increases by 2π in making the circuit of the path.

The gravitational fields that occur in nature are conservative. If this were not so, it would be possible to accumulate unlimited amounts of work by merely allowing particles to describe closed paths.

11·2. Virtual Work. From **11·12** an immediate deduction can be made, viz. (i) *If a system of forces acting on a particle is in equilibrium and the particle undergoes a small displacement, the algebraical sum of the work done by the forces is zero.* This follows from the fact that the work done by the forces is equal to the work done by their resultant, and this is zero because they have a zero resultant.

There is a converse to this proposition which can be proved

in the same way, viz. (ii) *If the algebraical sum of the work done by a system of forces acting on a particle is zero for all small displacements of the particle, then the forces are in equilibrium.* For, from **11·12** if the forces have a resultant, the work done by the resultant in any displacement is equal to the algebraical sum of the work done by the forces, i.e. equal to zero. But the work done by a force can only be zero if *either* the force is zero, *or* the displacement is at right angles to the direction of the force. So if the work done is zero, no matter what be the direction of the displacement, it follows that the resultant force must be zero and the given forces are in equilibrium.

These propositions are known as the Principle of Virtual Work and its Converse *for forces acting upon a single particle.*

The word *virtual* is used to qualify work, because we are dealing with statical systems in which there is no motion, and the displacements are hypothetical or virtual, so that the work calculated is the amount of work that would be done if the displacements were actually made.

11·21. Examples. (i) *A particle of weight W is supported on a smooth plane inclined at an angle α to the horizontal by a force P which makes an angle θ with the plane.*

Let R be the normal reaction of the plane on the particle. Consider a small displacement of the particle up the plane from A to A'. Since AA' is at right angles to the direction of R, therefore no work is done by R; but P does work $P.AA'\cos\theta$ and W does work $-W.AA'\sin\alpha$, and in equilibrium the total work done is zero, therefore

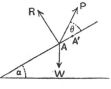

$$P.AA'\cos\theta - W.AA'\sin\alpha = 0,$$

or $$P\cos\theta = W\sin\alpha.$$

The application of the principle of virtual work in this case gives the same result as would have been obtained by resolving in the direction AA', as we might have anticipated from the nature of the proof of the principle. In fact in this case the principle of virtual work gives us no more information than we already possess.

Let us next consider an application of the principle to a case in which two particles are concerned.

(ii) *Two small rings can slide on a smooth parabolic wire with its axis vertical and vertex upwards, being connected by a fine string which passes over a smooth peg at the focus. Shew that if the rings can rest in any position in which the string is taut they must be of equal weight.*

Let W, W' be the weights of the rings at P and Q. Let S be the focus.

Each ring is supported by the tension of the string and the pressure of the wire, so that if the tensions of the two portions of the string PS, QS are equal there will be equilibrium. Consider a small displacement

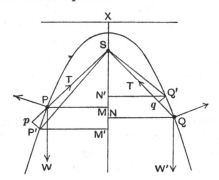

in which the rings move to P' and Q'. The normal reactions being at right angles to the displacements do no work. The work done by W at P is $W.MM'$, where M, M' are the projections of P, P' on the axis; and the work done by T at P is $-T.Pp$, where Pp is the projection of PP' on SP. Therefore, by applying the principle of virtual work to the ring at P, we get

$$W.MM' - T.Pp = 0 \quad \ldots\ldots\ldots\ldots\ldots\ldots(1).$$

Similarly by considering the ring at Q we get

$$-W'.N'N + T.Qq = 0 \quad \ldots\ldots\ldots\ldots\ldots(2).$$

Now by hypothesis $P'S + SQ' = PS + SQ$, and, to the first order of small quantities, $P'S = pS$ and $SQ' = Sq$, so that $Pp = Qq$, and by adding (1) and (2) we get

$$W.MM' - W'.N'N = 0 \quad \ldots\ldots\ldots\ldots\ldots(3).$$

But, if the axis meets the directrix in X, we have

$$MM' = XM' - XM = SP' - SP$$
$$= SQ - SQ'$$
$$= XN - XN' = N'N.$$

Therefore $W = W'$.

11·22. In the preceding example we found it necessary to introduce the work done by the mutual action and reaction between the particles, viz. the pull of each on the string connecting them, and we found that the tensions at opposite ends of the string contributed equal and opposite amounts to the *total* work done.

We now observe that in this case also our applications of the principle of virtual work to the rings separately, as expressed in equations (1) and (2), gave us no more information than we should have got by resolving along the tangents at P and Q; for, as appeared later, $MM' = Pp$ and $NN' = Qq$. But if we had chosen to write down in one equation the total work done by all the forces acting on the two rings and at the same time had been able to foresee that the work done by the mutual action and reaction along the string would cancel out, we should have written down at once equation (3), and the problem would have been solved without any reference to the tension of the string.

11·23. Our object is now to extend the Principle of Virtual Work, proved for a single particle in **11·2**, and to make it applicable, in a much more general way, to a body or system of bodies in equilibrium under the action of given *external* forces. In **11·22** we pointed out how we could have curtailed the work in **11·21** Ex. (ii) if we could have omitted the work done by a mutual action and reaction, and it is only because it is possible to omit from the work sum in the general case all the work done by internal actions and reactions and by forces of constraint (subject to certain limitations) that the principle of virtual work, which we are about to establish, has any practical value.

As the next step therefore in establishing the principle, we will shew that there are a number of cases in which mutual actions and reactions and forces of constraint either do equal and opposite amounts of work, or do no work, in a small displacement.

But we must first explain that the virtual displacements with which we are concerned are to be regarded as small quantities 'of the first order of smallness', and we shall consider that forces do 'no work' when the algebraical sum of the work done by them in terms of the displacements is 'of the second or higher order of smallness'.

(i) *The work done by the mutual action and reaction between two particles is zero if their distance apart is invariable.*

Let F be the force acting between the particles A, B which undergo a small displacement to A', B' so that $A'B' = AB$. Let θ be the small angle between $A'B'$ and AB, and let M, N be the projections of

A', B' upon AB. The work done by the two forces F

$$= F . AM - F . BN$$
$$= F (AB - MN) = F (AB - A'B' \cos \theta)$$
$$= F . AB (\tfrac{1}{2} \theta^2 - \text{higher powers of } \theta)$$
$$= 0, \text{ to the first order of the small displacement } \theta.$$

(ii) *The work done by the tensions at the ends of an inextensible string in contact with smooth surfaces is zero for a small displacement.*

Assuming that the string is acted upon by no forces save tensions at its ends and the normal pressures of smooth surfaces, if we consider the equilibrium of any small element of

the string the only forces upon it in the directions of its length are the tensions at its ends, so these must be equal and opposite and the tension is therefore constant throughout the whole length of the string.

Let $APQB$ represent the string in the equilibrium position, and suppose that the ends A, B are displaced to A', B', where APA', BQB' are small angles θ, ϕ. Since the string is inextensible
$$A'P + QB' = AP + QB.$$

If T denotes the tension at either end and M, N are the projections of A', B' on PA, BQ, the total work done

$$= T . AM - T . BN$$
$$= T (PM - PA) - T (BQ - NQ)$$
$$= T (PA' \cos \theta - PA) - T (BQ - B'Q \cos \phi)$$
$$= T (PA' - PA - BQ + B'Q), \text{ to the first order of } \theta, \phi$$
$$= 0.$$

(iii) *The reaction of a smooth surface upon a body which slides on the surface does no work;* because the reaction is always along the normal and therefore at right angles to the direction of the displacement of its point of action.

(iv) *The reaction of a fixed surface upon a body which rolls on the surface does no work.*

Let A be the point of contact, and after the moving body has rolled through a small angle θ let B be the point of contact and A' the new position of the point A on the rolling body.

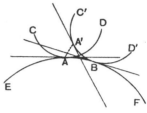

The work done is proportional to AA', and AA' is of order $AB.\theta$.

Now θ is the sum of the angles which the tangents at A and A' make with the tangent at B, therefore

$$\theta = \frac{AB}{\rho} + \frac{A'B}{\rho'},$$

where ρ, ρ' are the radii of curvature of the curves EAF, CAD; and $A'B = AB$, therefore AB is of order θ and AA' is of order θ^2, and the work done by the reaction at A is therefore negligible.

(v) *The mutual reaction between two bodies which roll upon one another does no work in a small displacement.*

In this case both bodies are displaced. Referring to the last figure, we may suppose the total displacement to be performed in two stages. Firstly, keeping the bodies fixed relatively to one another, let the body EAF move into its final position. During this part of the displacement the mutual action and reaction at A do equal and opposite amounts of work, because the points A on both the bodies are equally displaced. Secondly, keeping the body EAF fixed, we may allow the body CAD to roll into its final position, then, since the whole displacement is small, by (iv) no work is done by the reaction at A on CAD, and therefore the total work done by the reactions remains zero.

11·3. The Principle of Virtual Work for a System of Bodies. A body or a system of bodies may be regarded as an aggregate of material particles held together by forces of cohesion together with such mutual actions and reactions as we have classed in **3·6** as 'internal forces'; and in the most general case the possible displacements of such a body or system may be restricted by certain constraints of the type described in **3·7** and to which reference is made in **6·2**. The theorem proved in **11·2**, viz. that if a system of forces acting on a particle is in equilibrium and the particle undergoes a small displacement the algebraical sum of the work done by the forces is zero, is true for the forces acting upon all the particles that go to make up a body or system of bodies, so that by addition it follows that in *any* small displacement of a body or system of bodies in equilibrium the total work done by *all* the forces acting upon *all* the separate particles which compose the body or system is zero.

Now in the case of a single *rigid* body in equilibrium under the action of given external forces, there are no internal forces save such as come under the category of mutual action and reaction between particles whose distance apart is invariable, and by **11·23** (i) such forces contribute nothing to the total work done in a small displacement.

Therefore *when a single rigid body in equilibrium under the action of given external forces undergoes a small displacement the algebraical sum of the work done by the external forces is zero.*

Further, for a system of bodies in equilibrium under the action of given external forces and certain forces of constraint including the mutual actions and reactions between the bodies, we have seen in **11·23** (iii), (iv) and (v) certain types of relative displacement of bodies in which constraining forces or mutual actions and reactions do no work; consequently, if we restrict the small displacement of the system so that it only involves relative displacements of the types stated, then the total work done by all mutual actions and reactions and forces of constraint will amount to zero; but the total work done by *all* the forces acting upon the separate particles of the system is zero, and as the forces consist of the given external forces together with mutual actions and reactions and forces of constraint and, as we have just seen that the latter together contribute nothing to the work sum, therefore the total work done by the external forces alone must be zero.

We have thus arrived at the most general form of the Principle of Virtual Work, which we may state as follows:

If for a system of bodies in equilibrium under the action of given external forces and subject to certain constraints it is possible to make a displacement such that the constraining forces do no work, then, for any such displacement, the algebraical sum of the work done by the external forces alone is zero, or of a higher order of smallness than the first in terms of the displacements.

11·31. The Converse Theorem. In **11·2** we established a converse theorem for forces acting on a single particle, viz. that if the algebraical sum of the work done by the forces is zero for all small displacements of the particle, then the particle is in equilibrium. The converse of the general theorem is:

If a system of bodies is subject to the action of given external forces which are such that, for all small displacements in which the forces of constraint do no work, the algebraical sum of the work done by the external forces is zero, then the system of bodies is in equilibrium.

It is clear that this cannot be deduced by any process of summing the results for separate particles, because the data do not include any information about the forces which act upon separate particles; and it is not easy to give a satisfactory direct proof.

The following argument is open to the criticism that the assumption made is one of which it would be difficult to formulate an explicit proof:

Given that the algebraical sum of the work done by the external forces vanishes for all small displacements in which the forces of constraint do no work, to prove that the system on which the forces act is in equilibrium. It if were not in equilibrium it would move. Suppose that the system is surrounded by such frictionless constraints as will make the actual motion the only possible one, then we assume that this motion could be prevented by the application of a single force F, say, suitably applied at a point A. Let this force be applied, so that the system is now in equilibrium under the action of F and the given external forces and the smooth constraints. Apply the principle of virtual work taking a small displacement in the direction of the hypothetical motion. This will displace A to A' through a small distance $-df$ in the direction of F, and give an equation of virtual work

$-F.df+$ the work done by the external forces $=0$.

But by hypothesis the work done by the external forces alone is zero, and df is not zero, therefore F is zero and the system is in equilibrium.

Later in this chapter (11·5) we shall shew that the vanishing of the algebraical sum of the work done by the external forces acting on a single rigid body in arbitrarily chosen small displacements leads to the same equations of equilibrium as are obtained by resolving and taking moments, and that fact constitutes a justification for assuming the truth of the converse proposition as enunciated above for a single rigid body.

11·32. A special case. The following simple proof of the principle of virtual work and its converse for coplanar forces acting upon *a single rigid body* is due to the late Dr W. H. Besant.

Let P be any one of the forces acting at a point A. Let the body receive a small displacement so that the plane in which the forces act remains unchanged. Such a displacement must consist in a rotation of the body in this plane about an instantaneous centre* I through a small angle θ, whereby A is displaced to A'. Then if $A'M$ and IY are at right

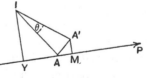

angles to the line of action of P, the work done by all the forces]

$$=\Sigma P.AM=\Sigma P.AA'\cos MAA'$$
$$=\Sigma P.IA.\theta\cos AIY$$
$$=\theta\Sigma P.IY$$
$$=\theta \times \text{algebraical sum of the moments of the forces about } I$$
$$=0, \text{ since the forces are in equilibrium.}$$

Conversely, if the algebraical sum of the work done by the forces is zero, no matter about what point in the plane the rotation of the body takes place, it follows that the algebraical sum of the moments of the forces about each such point is zero and the forces must be in equilibrium.

11·33. Applications. The most common direct application of the principle is to determine the position of a body or system of bodies in equilibrium under the action of given external forces, when the body or system has one or more degrees of

* *Dynamics*, 5·41.

freedom, such that small displacements are possible in which the constraining forces do no work.

In a large number of problems the only external forces are due to gravity, and the work done by the weights of the parts of a system in a small displacement is equal to the work done by the resultant weight acting at the centre of gravity of the system. But, by the principle of virtual work, no work is done by the external forces in a small displacement from a position of equilibrium, and this requires that, to the first order of small quantities, a small displacement of the system shall not produce any vertical displacement of its centre of gravity.

Hence, if z_1, z_2, z_3, ... denote the heights above a fixed level of the centres of gravity of the bodies of weights w_1, w_2, w_3, ... which are supported in a position of equilibrium, and $\bar{z}$ denotes the height of the centre of gravity of the whole, the principle of virtual work requires that $\bar{z}$ shall be constant for small displacements, or

$$d\bar{z} = 0 \quad \dots\dots\dots\dots\dots\dots\dots\dots(1).$$

But $w_1 z_1 + w_2 z_2 + w_3 z_3 + \dots = (w_1 + w_2 + w_3 + \dots)\,\bar{z},$

therefore the condition may also be expressed in the form

$$w_1 dz_1 + w_2 dz_2 + w_3 dz_3 + \dots = 0 \quad \dots\dots\dots\dots(2).$$

The position of the system is determined by one or more variables or co-ordinates of position θ, ϕ, ψ, ..., according to the number of degrees of freedom to be considered, so that z_1, z_2, z_3, ... are expressible in terms of θ, ϕ, ψ, ... and dz_1, dz_2, dz_3, ... are expressible in terms of $d\theta$, $d\phi$, $d\psi$, ...

After substitution, equation (2) therefore takes the form

$$\Theta\, d\theta + \Phi\, d\phi + \Psi\, d\psi + \dots = 0 \quad \dots\dots\dots\dots(3),$$

where Θ, Φ, Ψ, ... are functions of θ, ϕ, ψ, ..., and the number of these independent variables is the same as the number of degrees of freedom.

Since the variables are independent it is possible to choose a displacement in which, say, θ alone is varied while all the other variables ϕ, ψ, ... are kept constant; i.e. in (3) we may put

$$d\phi = d\psi = \dots = 0, \quad \text{while} \quad d\theta \neq 0,$$

so that from (3) we get $\Theta = 0.$

Similarly $\Phi = 0, \quad \Psi = 0, \text{ etc.}$

and in this way we get as many equations as there are variables and their solution determines the values of θ, ϕ, ψ, ... in the position of equilibrium.

Reverting to the original form of the condition as formulated in (1), we may also express it by saying that the height of the centre of gravity above (or depth below) a fixed level is a maximum or a minimum in a position of equilibrium. Also, since in geometry positions of symmetry are in general positions in which the variables that define the positions have maximum or minimum values, it is reasonable to expect that when a system of bodies and the forces acting upon them are capable of a symmetrical configuration it will be a position of equilibrium if there is one.

11·34. Examples with one degree of freedom. (i) *A rod lies in equilibrium with its ends on two smooth planes inclined at angles* α, β *to the horizontal, the planes intersecting in a horizontal line. Shew that the inclination of the rod to the horizontal is*

$$\tan^{-1}\frac{\sin(\alpha \sim \beta)}{2\sin\alpha\sin\beta}.$$ [S.]

It is evident that equilibrium is only possible when the vertical plane through the rod intersects the planes in lines of greatest slope. The rod has then one degree of freedom if it remains in contact with the planes, and its position is determined by its inclination θ to the horizontal.

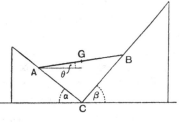

Let AB be the rod of length $2l$, and G its centre of gravity at a height z above the intersection C of the planes.

Then
$$z = l\sin\theta + AC\sin\alpha$$
$$= l\sin\theta + 2l\sin\alpha\sin(\beta-\theta)/\sin(\alpha+\beta),$$

and
$$dz = l\cos\theta\,d\theta - 2l\sin\alpha\frac{\cos(\beta-\theta)}{\sin(\alpha+\beta)}d\theta.$$

But in equilibrium $dz = 0$,

therefore $\sin(\alpha+\beta)\cos\theta - 2\sin\alpha\cos(\beta-\theta) = 0,$

which gives $\tan\theta = \dfrac{\sin(\beta-\alpha)}{2\sin\alpha\sin\beta}.$

(ii) *A heavy ring of radius r is supported by two thin smooth rods which pass through it and intersect at an angle 2α, the plane of the rods making an* *angle β with the horizontal. Prove that the depth of the centre of the ring below the point of intersection of the rods is*

$$r\sqrt{(1+\cot^2\alpha\sin^2\beta)}.$$ [S.]

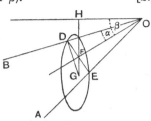

Let OA, OB be the rods, G the centre of the ring which touches the rods at D and E. Let the vertical through G meet the chord DE in F and the horizontal plane through O in H. Then assuming the position of equilibrium to be one in which $OD = OE$ and the plane of the ring is vertical, the angle

$$DOF = FOE = \alpha, \quad \text{and} \quad FOH = \beta.$$

Also, if $FO = x$, the depth of G below the level of O is

$$z = HF + FG = x\sin\beta + \sqrt{(r^2 - x^2\tan^2\alpha)},$$

therefore

$$dz = dx\sin\beta - \frac{x\,dx\,\tan^2\alpha}{\sqrt{(r^2 - x^2\tan^2\alpha)}}.$$

In equilibrium this expression vanishes, therefore

$$\sin\beta\sqrt{(r^2 - x^2\tan^2\alpha)} = x\tan^2\alpha,$$

or

$$\frac{r^2 - x^2\tan^2\alpha}{\tan^2\alpha} = \frac{x^2\tan^2\alpha}{\sin^2\beta} = \frac{r^2}{\tan^2\alpha + \sin^2\beta}.$$

Hence

$$z = \frac{r}{\sqrt{(\tan^2\alpha + \sin^2\beta)}}\left(\frac{\sin^2\beta}{\tan\alpha} + \tan\alpha\right)$$
$$= r\sqrt{(1 + \cot^2\alpha\sin^2\beta)}.$$

(iii) *A solid hemisphere is supported by a string fixed to a point on its rim and to a point on a smooth vertical wall with which the curved surface of the hemisphere is in contact. If θ, φ are the inclinations of the string and the plane base of the hemisphere to the vertical, prove that*

$$\tan\phi = \tfrac{3}{8} + \tan\theta.$$ [S.]

Let O be the point of suspension, AB the base of the hemisphere, C its centre and G its centre of gravity, OA the string and D the point of contact of the hemisphere and the wall.

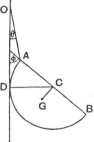

Let $OA = l$, and $CA = a$, then $CG = \tfrac{3}{8}a$, and the depth of G below the fixed point O is

$$z = l\cos\theta + a\cos\phi + \tfrac{3}{8}a\sin\phi,$$

so that $dz = -l\sin\theta\,d\theta + a(\tfrac{3}{8}\cos\phi - \sin\phi)\,d\phi$,

and this must vanish in the equilibrium position.

Now the variables $θ$, $φ$ are not independent because the radius CD is the horizontal projection of CA and AO, i.e.

$$a = a\sin\phi + l\sin\theta,$$

so that

$$a\cos\phi\,d\phi + l\cos\theta\,d\theta = 0;$$

but we also have

$$-l\sin\theta\,d\theta + a\,(\tfrac{3}{8}\cos\phi - \sin\phi)\,d\phi = 0,$$

and, eliminating the ratio $d\theta : d\phi$. we find that

$$\tan\phi = \tfrac{3}{8} + \tan\theta.$$

11·35. The following examples will serve to illustrate a method of using the principle of virtual work to determine an unknown force or couple necessary to maintain a given position of equilibrium:

(i) *A rod AB is moveable about a pivot at A, and to B is attached a string whose other end is tied to a ring. The ring slides along a smooth horizontal wire passing through A; prove by the principle of work that the horizontal force necessary to keep the ring at rest is $\dfrac{W\cos\alpha\cos\beta}{2\sin(\alpha+\beta)}$, where W is the weight of the rod and α, β the inclinations of the rod and string to the horizontal.* [S.]

Let C be the ring and P the horizontal force.

Let $AB = a$, $BC = l$ and $AC = x$.

Consider a small displacement in which a and l remain invariable while the angles change. The equation of virtual work is

$$P\,dx + W\,d\,(\tfrac{1}{2}a\sin\alpha) = 0 \quad \ldots(1).$$

But $\qquad x = a\cos\alpha + l\cos\beta,$

so that (1) is equivalent to

$$-P\,(a\sin\alpha\,d\alpha + l\sin\beta\,d\beta) + \tfrac{1}{2}Wa\cos\alpha\,d\alpha = 0 \quad \ldots\ldots(2)$$

Now α and β are not independent but connected by the relation

$$a\sin\alpha = l\sin\beta,$$

so that $\qquad a\cos\alpha\,d\alpha = l\cos\beta\,d\beta \quad \ldots\ldots\ldots\ldots\ldots(3).$

Then by eliminating the ratio $d\alpha : d\beta$ from (2) and (3) we get

$$P\,(\tan\alpha + \tan\beta) = \tfrac{1}{2}W,$$

or $\qquad P = W\cos\alpha\cos\beta/2\sin(\alpha+\beta).$

(ii) *Example to illustrate work done by a couple.*

Let P, P be the forces of the couple acting at the end of the arm AB. It is clear that the two forces will do equal and opposite amounts of work in any general translational displacement of the arm AB that does not involve rotation. Also no work will be done in displacement of the arm AB in a plane at right angles to the direction of the forces. Hence work is only done when the arm rotates in the plane of the couple. Also it is im-

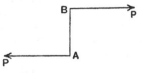

material about what point the arm rotates; because by a translation, involving no work, any chosen point of the arm can be brought into coincidence with any point in space. Consider then a rotation of the arm AB about A through a small angle $d\theta$; B receives a small displacement $BB' = AB . d\theta$ and the work done by the couple

$$= P . AB . d\theta$$

= the moment of the couple × the angle turned through.

By addition the same result is true for displacement through an angle of any magnitude.

A uniform square lamina of side a and weight W is suspended from four points in a horizontal plane by equal inextensible vertical strings of length l attached to the corners of the square. Shew that the couple that would be necessary to hold the lamina in a position in which it has been turned through an angle θ from the former position is

$$Wa^2 \sin\theta / 2\sqrt{\{l^2 - a^2(1 - \cos\theta)\}}.$$

Let O be the centre of the lamina $ABCD$ before the couple is applied and AP, BQ, CR, DS the strings. Suppose that O', A' are the displaced positions of O, A when a couple L is acting on the lamina. Let $O'M$ be the perpendicular from O' to AP. Then

$$A'O'M = \theta, \quad A'P = AP = l,$$

and $O'M = O'A' = OA = a/\sqrt{2}.$

If $OO' = z$, the equation of virtual work for a further small displacement is $Ld\theta - Wdz = 0.$

But $A'P^2 = A'M^2 + MP^2,$

or $l^2 = 2a^2 \sin^2 \tfrac{1}{2}\theta + (l - z)^2.$

Whence we get $0 = a^2 \sin\theta \, d\theta - 2(l - z)\,dz,$

so that $$\frac{L}{a^2 \sin\theta} = \frac{W}{2(l - z)},$$

or $L = Wa^2 \sin\theta / 2\sqrt{\{l^2 - a^2(1 - \cos\theta)\}}.$

11·36. Examples with two degrees of freedom. (i) *Weights w_1, w_2 are fastened to a light inextensible string ABC at the points B, C, the end A being fixed. Prove that, if a horizontal force P is applied at C and in equilibrium AB, BC are inclined at angles θ, ϕ to the vertical, then*

$$P = (w_1 + w_2)\tan\theta = w_2 \tan\phi.$$

Let $AB = a$ and $BC = b$.

Here there are two independent variables θ, ϕ. A is the only fixed point, and the horizontal distance of C from the vertical through A is

$$a\sin\theta + b\sin\phi,$$

so that the work done by P in a small displacement is
$$Pd\,(a\sin\theta+b\sin\phi),$$
and the equation of virtual work is
$$Pd\,(a\sin\theta+b\sin\phi)+w_1d\,(a\cos\theta)$$
$$+w_2d\,(a\cos\theta+b\cos\phi)=0,$$
or $\quad(Pa\cos\theta-w_1a\sin\theta-w_2a\sin\theta)\,d\theta$
$$+(Pb\cos\phi-w_2b\sin\phi)\,d\phi=0.$$

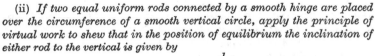

It will be observed that this equation is
of the type (3) of **11·33**, and that since θ
and ϕ are independent so are $d\theta$ and $d\phi$;
i.e. the small values of $d\theta$ and $d\phi$ may be
chosen arbitrarily, so that we may take
$d\phi=0$ and $d\theta\neq0$, giving
$$P=(w_1+w_2)\tan\theta,$$
or $d\theta=0$ and $d\phi\neq0$, giving
$$P=w_2\tan\phi.$$

(ii) *If two equal uniform rods connected by a smooth hinge are placed
over the circumference of a smooth vertical circle, apply the principle of
virtual work to shew that in the position of equilibrium the inclination of
either rod to the vertical is given by*
$$\operatorname{cosec}^2\theta\cot\theta=\frac{l}{a},$$
*where $2l$ and $2a$ are the length of a rod and the diameter of the circle
respectively.*   [S.]

If we assume that the position of equilibrium is necessarily a symmetrical one, then there is only one
degree of freedom; the point of intersection A of the rods AB, AC being
free to move up and down the vertical
through the centre of the circle. But
we can prove that a position of equilibrium *must* be a symmetrical one
by allowing that two degrees of freedom are possible, and taking 2θ to
denote the angle between the rods
and ϕ to denote the inclination of
OA to the vertical.

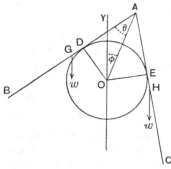

Let G, H be the centres of gravity
of the rods AB, AC and D, E the
points of contact of the rods with the circle.
Then $AD=AE=a\cot\theta$, and the height of G above O is
$$OD\sin(\theta+\phi)-DG\cos(\theta+\phi),$$
or $\qquad\qquad a\sin(\theta+\phi)-(l-a\cot\theta)\cos(\theta+\phi).$
Similarly the height of H above O is
$$a\sin(\theta-\phi)-(l-a\cot\theta)\cos(\theta-\phi).$$

The height above O of the centres of gravity of the two rods is half of the sum of the heights of G and H, i.e.

$$\bar{z} = a\sin\theta\cos\phi - (l - a\cot\theta)\cos\theta\cos\phi$$
$$= a\,\mathrm{cosec}\,\theta\cos\phi - l\cos\theta\cos\phi,$$

and in equilibrium $\qquad d\bar{z} = 0,$
therefore

$$-(a\,\mathrm{cosec}\,\theta - l\cos\theta)\sin\phi\,d\phi$$
$$-(a\,\mathrm{cosec}\,\theta\cot\theta - l\sin\theta)\cos\phi\,d\theta = 0 \ \dots\dots(1).$$

Now θ and ϕ, and therefore also $d\theta$ and $d\phi$, are independent. If we put $d\theta = 0$ and $d\phi \neq 0$, we must have

$$(a\,\mathrm{cosec}\,\theta - l\cos\theta)\sin\phi = 0 \ \dots\dots\dots\dots(2);$$

and one solution is $\phi = 0$, which gives a position of symmetry.
Then putting $d\phi = 0$ and $d\theta \neq 0$ in (1), we get

$$(a\,\mathrm{cosec}\,\theta\cot\theta - l\sin\theta)\cos\phi = 0 \ \dots\dots\dots\dots(3),$$

and in the symmetrical position this gives

$$\mathrm{cosec}^2\,\theta\cot\theta = \frac{l}{a} \ \dots\dots\dots\dots\dots\dots(4).$$

The alternative solution of (2) is

$$\sin\theta\cos\theta = \frac{a}{l}, \quad \phi \neq 0 \dots\dots\dots\dots\dots(5).$$

But we have to combine this with (3), and this implies that either $\phi = \frac{1}{2}\pi$ or $\mathrm{cosec}^2\,\theta\cot\theta = l/a$.

Now it is easy to see that, if the angle YOE were greater than $\frac{1}{2}\pi$, contact between the rod AC and the circle would be broken; so that we may reject $\phi = \frac{1}{2}\pi$ as impossible, and there remains the combination of (5) and $\mathrm{cosec}^2\,\theta\cot\theta = \dfrac{l}{a}$.

This leads to $\theta = \frac{1}{4}\pi$ and $l = 2a$, and refers to a special case in which a possible position of equilibrium is with one rod horizontal and the figure $ADOE$ a square; and this is only possible when $l = 2a$.

Hence in general the position of equilibrium must be a symmetrical one in which the inclination of the rods to the vertical is given by (4).

11·4. Application of the Principle of Virtual Work to the determination of Unknown Reactions. The second class of problem to which the principle of virtual work is applicable is one in which the equilibrium configuration of a system under the action of given external forces is known and it is required to determine one or more of the internal actions and reactions. In establishing the principle in its general form in **11·3** we imposed a restriction on the displacements for which the algebraical sum of the work done by the external forces alone is zero, viz. that any such displacement must be one in which the constraining forces do no work; but it is evident from the proof of the principles that, if the displacement chosen be such that some one constraining force does work, the theorem would be valid for that displacement also provided that we add to the work done by the external forces the work done by this particular constraining force. Then equating to zero the total work done we have an equation in which this particular constraining force appears as the only unknown quantity and is therefore determined in this way.

A similar argument applies to an unknown action and reaction between two particles provided the distance between them does not remain constant during the displacement, for in this case they make a contribution to the total work done and appear as an unknown quantity in the equation of virtual work.

The method will easily be understood after the working of a few examples.

11·41. In many cases it is required to find the tension or thrust in a light rod joining two points A, B of a framework.

If the rod is rigid, it is not possible to make a displacement which alters the distance AB. But if T is the tension in the rod, the equations of equilibrium of the surrounding bodies are unaltered if we replace the rod by a force T acting at A along AB and a force T acting at B along BA. We can now make a displacement of the system which alters the distance AB, say, increases it from l to $l+dl$.

Then if A', B' denote the displaced positions of A and B and M, N their projections on AB, the work done by the tension

$$= T.AM - T.BN$$
$$= T(AB - MN)$$
$$= T(AB - A'B'),$$

to the first order of the small angle between $A'B'$ and AB,
$$= -Tdl.$$

Similarly the work done by a thrust T in a virtual extension of a rod from length l to length $l+dl$ is $+Tdl$.

In this way we see that it is not necessary to consider the separate virtual displacements of the ends of the rod, but merely its increment in length.

11·42. Examples. (i) *A quadrilateral $ABCD$ formed of four uniform rods freely jointed to each other at their ends, the rods AB, AD being equal* *and also the rods BC, CD, is freely suspended from the joint A. A string joins A to C and is such that ABC is a right angle. Apply the principle of virtual work to shew that the tension of the string is* $(w+w')\sin^2\theta+w'$, *where w is the weight of an upper rod and w' of a lower rod and* 2θ *is equal to the angle BAD.* [S.]

Let $AB=AD=2a$,

and $CB=CD=2b$.

Let the angle $BCD=2\phi$.

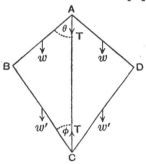

In the given position of equilibrium ϕ is the complement of θ, ABC being a right angle; but if at the outset we write $\frac{1}{2}\pi-\theta$ instead of ϕ, this implies a relation

$$b=a\tan\theta$$

and it would be impossible to make an increment in θ without altering b or a. This would introduce into the equation of virtual work the unknown stresses in the sides of the quadrilateral with which we are not concerned. We therefore use an independent symbol for the angle ACB, and imagine a displacement which alters the angles of the figure and the length AC but not the lengths of the sides of the quadrilateral.

The equation of virtual work is then

$$-Td(2a\cos\theta+2b\cos\phi)$$
$$+2wd(a\cos\theta)+2w'd(2a\cos\theta+b\cos\phi)=0,$$

or $T(a\sin\theta d\theta+b\sin\phi d\phi)$
$$-wa\sin\theta d\theta-w'(2a\sin\theta d\theta+b\sin\phi d\phi)=0.$$

But θ and ϕ are connected by

$$a\sin\theta=b\sin\phi,$$

so that $a\cos\theta d\theta=b\cos\phi d\phi$;

and, dividing the equation of virtual work by either of these equal expressions, we get

$$T(\tan\theta+\tan\phi)=w\tan\theta+w'(2\tan\theta+\tan\phi).$$

Now put $\cot\theta$ instead of $\tan\phi$, and we find that

$$T=(w+w')\sin^2\theta+w'.$$

(ii) *A smoothly jointed framework of light rods forms a quadrilateral ABCD. The middle points P, Q of an opposite pair of rods are connected by a string in a state of tension T, and the middle points R, S of the other pair by a light rod in state of thrust X; shew by the method of virtual work that* $T/PQ=X/RS$. [S.]

The equation of virtual work for any small displacement which alters the angles but not the lengths of the sides of the quadrilateral $ABCD$ is $-Td(PQ)+Xd(RS)=0$ ....(1),

in accordance with the explanation given in **11·41**.

It is necessary therefore to find a relation between PQ and RS.

PR and QS are both parallel to BD, and PS and RQ are both parallel to AC, therefore $PRQS$ is a parallelogram and therefore PQ and RS bisect one another in O.

Then since OQ is a median of the triangle BOC,

$$BO^2 + OC^2 = \tfrac{1}{2}PQ^2 + \tfrac{1}{2}BC^2;$$

similarly

$$CO^2 + OD^2 = \tfrac{1}{2}RS^2 + \tfrac{1}{2}CD^2.$$

Therefore $\quad 2(BO^2 - OD^2) = PQ^2 - RS^2 + BC^2 - CD^2.$

Similarly $\quad 2(BO^2 - OD^2) = RS^2 - PQ^2 + AB^2 - DA^2;$

so that by subtraction

$$2(PQ^2 - RS^2) = AB^2 - BC^2 + CD^2 - DA^2.$$

Now in our hypothetical displacement the lengths of PQ, RS are altered but not the lengths of the sides of the quadrilateral, therefore

$$PQ \cdot d(PQ) - RS \cdot d(RS) = 0 \quad \dots\dots\dots\dots(2);$$

and from (1) and (2) it follows that

$$T/PQ = X/RS.$$

(iii) *$ABCDEF$ is a regular hexagon formed of light rods smoothly jointed at their ends with a diagonal rod AD. Four equal forces P act inwards at the middle points of the rods AB, CD, DE, FA and at right angles to the respective sides. Find the stress in the diagonal AD and state whether it is a tension or a thrust.*

Let O, H, K, L, M be the middle points of AD, AB, CD, DE, FA.

Let T be the stress in AD assumed to be a thrust.

We must not at the outset assume that the hexagon is regular, for we want to be able to make a displacement which will alter the length of AD but leave the lengths of the sides of the hexagon unaltered. Hence we suppose that the figure is symmetrical about AD and denote the angles BAD, FAD, CDA and EDA by θ.

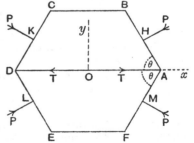

Then if $2a$ denotes the length of a side of the hexagon, the length of AD is $2a(1 + 2\cos\theta)$, and the work done by the thrust T in a small displacement which alters the length of AD is

$$Td[2a(1 + 2\cos\theta)],$$

or $\qquad\qquad -4aT\sin\theta\,d\theta.$

For the sake of simplicity we can suppose that the displacement does not alter the position of the point O, but merely alters the position of A and D in the line AD, with a consequent change in θ and in the positions of the points H, K, L, M. By symmetry the forces P at these four points will contribute equal amounts of work, so that it will suffice to determine the work done by P at H.

Take an axis Ox along OA and a perpendicular axis Oy as in the figure. The co-ordinates of H are

$$x = a + a\cos\theta, \quad y = a\sin\theta,$$

and the components of P at H are

$$X = -P\sin\theta, \quad Y = -P\cos\theta.$$

The work done by P in a small displacement is therefore

$$X\,dx + Y\,dy = +Pa\sin^2\theta\,d\theta - Pa\cos^2\theta\,d\theta$$

$$= -Pa\cos 2\theta\,d\theta.$$

Hence the equation of virtual work is

$$-4aT\sin\theta\,d\theta - 4Pa\cos 2\theta\,d\theta = 0.$$

And for a regular hexagon $\theta = \tfrac{1}{3}\pi$, so that

$$T = P/\sqrt{3},$$

and since the sign of T is positive the stress is a thrust.

11·5. Deduction of the Conditions of Equilibrium of a Rigid Body from the Principle of Virtual Work.

(i) *When the forces are coplanar.* The most general small displacement of the body in the plane of the forces consists of a translation and a rotation.

Take a set of rectangular axes Ox, Oy in the plane of the forces. Let A be the point of the body at which a force P is applied, and let P_x, P_y denote the components of P parallel to the axes and x, y the co-ordinates of A, so that $OM = x$ and $MA = y$. Consider the effect on the diagram of a small rotation of the body about O through an angle $\delta\theta$ in the sense from Ox towards Oy, the axes remaining fixed in space.

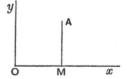

The displacement of A relative to M would be $y\,\delta\theta$ parallel to MO, and the displacement of M would be $x\,\delta\theta$ along MA. Hence the rotation decreases the x co-ordinate of A by $y\,\delta\theta$ and increases the y co-ordinate by $x\,\delta\theta$ in reference to the original position of the axes.

If in addition the body undergoes a small translation whose

components parallel to the axes are δx and δy, then the *total* displacements of the point A parallel to the axes are

$$dx = \delta x - y\,\delta\theta \quad \text{and} \quad dy = \delta y + x\,\delta\theta.$$

Hence, if the body is in equilibrium under the action of a set of forces of which P is a type, we have an equation of virtual work

$$\Sigma\,(P_x dx + P_y dy) = 0,$$

or
$$\delta x.\Sigma P_x + \delta y.\Sigma P_y + \delta\theta.\Sigma\,(x P_y - y P_x) = 0.$$

But the displacements δx, δy, $\delta\theta$ are independent and arbitrary, therefore

$$\Sigma P_x = 0, \quad \Sigma P_y = 0$$

and
$$\Sigma\,(x P_y - y P_x) = 0.$$

These equations represent the ordinary conditions of equilibrium, viz. that the algebraical sums of the resolved parts of the forces in any two directions at right angles are zero, and the algebraical sum of the moments of the forces about a point in the plane is zero.

(ii) *When the forces are not coplanar.* We now take a set of three rectangular axes Ox, Oy, Oz. Let A be the point of the body at which a force P is applied whose components parallel to the axes are P_x, P_y, P_z.

Let x, y, z be the co-ordinates of A, where in the figure $OL = x$, $LM = y$ and $MA = z$. The most general small rotational displacement of the body is compounded of

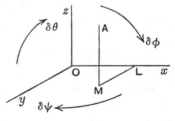

a small rotation $\delta\theta$ about Ox,

a small rotation $\delta\phi$ about Oy,

and a small rotation $\delta\psi$ about Oz,

in the senses indicated by the arrows.

Considering the effect on the diagram of these rotations, the displacements of A relative to M would be

$$z\,\delta\theta \text{ parallel to } yO \quad \text{and} \quad z\,\delta\phi \text{ parallel to } Ox;$$

the displacements of M relative to L would be

$$y\,\delta\psi \text{ parallel to } xO \quad \text{and} \quad y\,\delta\theta \text{ parallel to } Oz;$$

and the displacements of L relative to O would be

$$x\,\delta\phi \text{ parallel to } zO \quad \text{and} \quad x\,\delta\psi \text{ parallel to } Oy.$$

Hence the combined effect of the rotations is to increase the x, y and z co-ordinates of A in reference to the original position of the axes by

$$z\,\delta\phi - y\,\delta\psi, \quad x\,\delta\psi - z\,\delta\theta \quad \text{and} \quad y\,\delta\theta - x\,\delta\phi$$

respectively.

If in addition the body undergoes a small translation whose components parallel to the axes are δx, δy and δz, then the total displacements of the point A parallel to the axes are given by

$$dx = \delta x + z\,\delta\phi - y\,\delta\psi,$$
$$dy = \delta y + x\,\delta\psi - z\,\delta\theta,$$
and
$$dz = \delta z + y\,\delta\theta - x\,\delta\phi.$$

Hence, if the body is in equilibrium under the action of a set of forces of which P is a type, we have an equation of virtual work

$$\Sigma(P_x dx + P_y dy + P_z dz) = 0,$$

or
$$\delta x.\Sigma P_x + \delta y.\Sigma P_y + \delta z.\Sigma P_z$$
$$+ \delta\theta.\Sigma(yP_z - zP_y) + \delta\phi.\Sigma(zP_x - xP_z) + \delta\psi.\Sigma(xP_y - yP_x) = 0.$$

But the displacements δx, δy, δz, $\delta\theta$, $\delta\phi$, $\delta\psi$ are independent and arbitrary, therefore

$$\Sigma P_x = 0, \quad \Sigma P_y = 0, \quad \Sigma P_z = 0,$$
and
$$\Sigma(yP_z - zP_y) = 0, \quad \Sigma(zP_x - xP_z) = 0, \quad \Sigma(xP_y - yP_x) = 0.$$

These equations express the facts that the algebraical sums of the resolved parts of the forces parallel to the axes and of the moments of the forces about the axes are separately zero. (See **14·2**.)

We have thus demonstrated that the assumption of the truth of the principle of virtual work leads to the ordinary equations of equilibrium; and we have also demonstrated the converse theorem for forces acting on a rigid body, because we have shewn that the vanishing of the work sum for all displacements implies that the conditions of equilibrium are satisfied.

11·6. Potential Energy. The potential energy of a body or system of bodies in a given configuration in a given field of force is the work that would be done by the forces acting on the body or system of bodies if the body or system moved from the given configuration to some standard configuration.

For example, considering the Earth's gravitational field, the weights of bodies are forces which are capable of doing work as the bodies descend from higher to lower levels. If we choose to regard the floor of the room as the level of zero potential energy, then a body at a higher level possesses an amount of potential energy measured by the work which the weight of the body would do if the body descended to the floor, i.e. measured by the weight of the body multiplied by the height of its centre of gravity above the floor.

It is not the absolute amount of potential energy possessed by a body that is important, but the *change* in its potential energy that would result from a given displacement. In the case just considered we might equally well choose the ceiling of the room as the level of zero potential energy and any body below the ceiling would then possess negative potential energy, equal to the work that would have to be done against its weight to raise it to the ceiling.

It follows from the definition of potential energy that when the forces acting on a system of bodies do positive work the potential energy of the system decreases by the amount of work done.

11·61. In **11·14** we defined the work function W or

$$\int (X\,dx + Y\,dy + Z\,dz)$$

for forces acting on a particle, and a similar definition holds for the work function of a system of forces acting on a body or system of bodies, and by analogy from **11·33** it follows that if $\theta, \phi, \psi, \ldots$ are a set of variables equal in number to the degrees of freedom of a system and defining its position and W is the work function, then dW can be expressed in the form

$$\Theta\,d\theta + \Phi\,d\phi + \Psi\,d\psi + \ldots,$$

this expression representing the work done in a small displacement of a general type.

But this work done would represent the corresponding loss of potential energy, so that, if V denotes the potential energy,

$$-dV = \Theta\,d\theta + \Phi\,d\phi + \Psi\,d\psi + \ldots \qquad \ldots\ldots\ldots(1).$$

But when V is a function of $\theta, \phi, \psi, \ldots$, we have

$$dV = \frac{\partial V}{\partial \theta}d\theta + \frac{\partial V}{\partial \phi}d\phi + \frac{\partial V}{\partial \psi}d\psi + \ldots \qquad \ldots\ldots\ldots(2).$$

Hence, since $d\theta, d\phi, d\psi, \ldots$ are independent, by comparing (1) and (2), we have

$$\Theta = -\frac{\partial V}{\partial \theta}, \quad \Phi = -\frac{\partial V}{\partial \phi}, \quad \Psi = -\frac{\partial V}{\partial \psi} \qquad \ldots\ldots\ldots(3).$$

Now just as in the expression

$$\int (X\,dx + Y\,dy + Z\,dz)$$

for the work function X, Y, Z denote components of force, so, in the expression (1), $\Theta, \Phi, \Psi, \ldots$ may be called *generalized components of force.* In particular for a displacement which consists of an increment $d\theta$ in θ, while the other variables ϕ, ψ, ... are unaltered, the work done, or decrease in potential energy is from (1) $-dV = \Theta \, d\theta,$

so that we may call Θ the 'force tending to increase θ', and from (3) we see that this force is $-\partial V/\partial \theta$.

If θ denotes a length, then Θ will be a force in the strict meaning of the word; but if θ denotes an angle, then Θ will be a couple tending to increase the angle.

11·62. The Energy Test of Stability. The dynamical principle of conservation of energy is that the sum of the potential and kinetic energies of a dynamical system is constant. It follows from this principle that whenever a system begins to move, since in the initial stages of the motion it acquires kinetic energy, therefore it loses potential energy.

This fact leads to a simple proof of the theorem that

Positions of maximum potential energy are positions of unstable equilibrium, and positions of minimum potential energy are positions of stable equilibrium.

Firstly, if the potential energy is stationary in value for small displacements of a system from a given position, no work is done in a small displacement, and by the converse of the principle of virtual work the system is in equilibrium.

Secondly, if the system is slightly displaced from a position of maximum potential energy and then set free, since it moves so that the potential energy decreases, therefore it moves farther away from the position of maximum potential energy, which must therefore be an unstable position.

If, on the other hand, the system is slightly displaced from a position of minimum potential energy and then set free, since it moves so that the potential energy decreases, therefore in this case it will move back towards the position of minimum potential energy, which must therefore be a stable position.

11·621. A simple illustration of the theorem of **11·62** is the case of a sphere free to move on a corrugated surface.

The weight of the sphere is the only force which does work in a displacement. Unstable positions are at the tops of the ridges as at A or C, where the potential energy measured by the height above a fixed level has maximum values, and stable positions are at the bottoms of the troughs as at B or D, where the potential energy has minimum values.

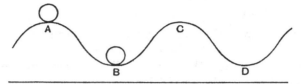

11·63. Applications. When the only external forces acting on a system are the weights of the bodies which compose it, the potential energy is measured by the whole weight multiplied by the height of the centre of gravity above a fixed level. This height $\bar{z}$ is expressible in terms of one or more variables according to the number of degrees of freedom of the system, and the positions of equilibrium are determined and their stability investigated by finding the values of the variables which give to $\bar{z}$ maximum and minimum values according to the ordinary methods of the differential calculus.

11·64. Examples. (i) *Two equal particles are connected by a light string which is slung over the top of a smooth vertical circle; verify that the position of equilibrium is unstable. (It may be supposed that both particles rest on the circle, so that the length of the string is less than one-half of the circumference of the circle.)* [S.]

Let the string be of length $2a\alpha$, where a is the radius of the circle, and let the radii drawn to the particles make angles θ, $2\alpha - \theta$ with the vertical.

The height of the centre of gravity of the two particles is given by

$$2\bar{z} = a\cos\theta + a\cos(2\alpha - \theta).$$

Therefore

$$\frac{d\bar{z}}{d\theta} = -a\sin\theta + a\sin(2\alpha - \theta),$$

and

$$\frac{d^2\bar{z}}{d\theta^2} = -a\cos\theta - a\cos(2\alpha - \theta).$$

The symmetrical position in which $\theta = \alpha$ is one in which $\dfrac{d\bar{z}}{d\theta} = 0$ and $\dfrac{d^2\bar{z}}{d\theta^2}$ is negative, i.e. a position in which $\bar{z}$ is a maximum, i.e. an unstable position of equilibrium.

(ii) *A solid circular cone whose height is h and semi-vertical angle α is placed vertex downwards in a smooth circular hole whose radius is a, cut* *in a horizontal table. Investigate the stability of possible positions of equilibrium.*   [S.]

Let A, B be the points at which the cone touches the table when its axis makes an angle θ with the vertical. Let O be the vertex, G the centre of gravity and M, N the projections of O and G on the plane of the hole.

Then $\bar{z} = NG = OG\cos\theta - OM$.

But

$$OM \cdot AB = 2\,(\text{area }OAB) = OA \cdot OB\sin 2\alpha$$
$$= OM^2\sec(\theta+\alpha)\sec(\theta-\alpha)\sin 2\alpha;$$

therefore

$$OM = 2a\cos(\theta+\alpha)\cos(\theta-\alpha)\operatorname{cosec}2\alpha$$
$$= a\,(\cos 2\theta + \cos 2\alpha)\operatorname{cosec}2\alpha,$$

and

$$\bar{z} = \tfrac{3}{4}h\cos\theta - a\,(\cos 2\theta + \cos 2\alpha)\operatorname{cosec}2\alpha.$$

Hence
$$\frac{d\bar{z}}{d\theta} = -\tfrac{3}{4}h\sin\theta + 2a\sin 2\theta\operatorname{cosec}2\alpha\dots\dots\dots\dots(1),$$

and
$$\frac{d^2\bar{z}}{d\theta^2} = -\tfrac{3}{4}h\cos\theta + 4a\cos 2\theta\operatorname{cosec}2\alpha\ \dots\dots\dots(2).$$

From (1) it follows that $d\bar{z}/d\theta$ vanishes for $\theta = 0$ and for

$$\cos\theta = \frac{3h}{8a}\sin\alpha\cos\alpha.$$

The value $\theta = 0$ refers to the symmetrical position of equilibrium, and for this value of θ we have from (2)

$$\frac{d^2\bar{z}}{d\theta^2} = -\frac{3h}{4} + 4a\operatorname{cosec}2\alpha.$$

This is positive if $8a > 3h\sin\alpha\cos\alpha$ and in this case $\bar{z}$ is a minimum and the symmetrical position is stable. In this case also the expression

$$\cos\theta = \frac{3h}{8a}\sin\alpha\cos\alpha$$

gives real values for θ on either side of the vertical, so that there are unsymmetrical positions of equilibrium. Further, when we substitute this value for $\cos\theta$ in (2), we find that

$$\frac{d^2\bar{z}}{d\theta^2} = 4a\left(\frac{9h^2}{64a^2}\sin^2\alpha\cos^2\alpha - 1\right)\operatorname{cosec}2\alpha;$$

and, if $8a > 3h\sin\alpha\cos\alpha$, this expression for $d^2\bar{z}/d\theta^2$ is negative, so that $\bar{z}$ is a maximum and the unsymmetrical positions are unstable. This fact could have been foreseen because maximum and minimum values of a function occur alternately, so that if the central symmetrical position is stable, the unsymmetrical positions of equilibrium on either side of the central one must be unstable.

If on the other hand $8a < 3h \sin \alpha \cos \alpha$, we find that the symmetrical position is unstable because $d^2\bar{z}/d\theta^2$ is then negative for $\theta = 0$; and in this case $\cos\theta = \dfrac{3h}{8a} \sin\alpha\cos\alpha$ does not give real values for θ, so that there are no unsymmetrical positions of equilibrium.

Finally, when $8a = 3h\sin\alpha\cos\alpha$, then we find that

$$\frac{d\bar{z}}{d\theta} = -4a\sin\theta\,(1 - \cos\theta)\operatorname{cosec} 2\alpha,$$

and this vanishes only for the symmetrical position in which $\theta = 0$.

Differentiating again, we find that

$$\frac{d^2\bar{z}}{d\theta^2} = -4a\operatorname{cosec} 2\alpha\,(\cos\theta - \cos 2\theta),$$

$$\frac{d^3\bar{z}}{d\theta^3} = 4a\operatorname{cosec} 2\alpha\,(\sin\theta - 2\sin 2\theta),$$

and $$\frac{d^4\bar{z}}{d\theta^4} = 4a\operatorname{cosec} 2\alpha\,(\cos\theta - 4\cos 2\theta).$$

Thus, for $\theta = 0$, the first derivative which does not vanish is of even order (the fourth) and negative, so that $\bar{z}$ is a maximum and the equilibrium is in this case also unstable.

(iii) *One end A of a uniform rod AB of weight W and length l is smoothly hinged at a fixed point, while B is tied to a light string which passes over a small smooth pulley a distance a vertically above A and carries a weight $W/4$. If $l < a < 2l$, shew that the system is in stable equilibrium when AB is vertically upwards, and that there is also a configuration of equilibrium in which the rod is inclined at a certain angle to the vertical.* [S.]

Let C be the pulley and b the length of the string, so that the portion of it hanging vertically is

$$b - BC = b - \sqrt{(a^2 + l^2 - 2al\cos\theta)}$$

when the rod AB makes an angle θ with the vertical AC.

The height $\bar{z}$ of the centre of gravity is given by

$$\tfrac{5}{4}W\bar{z} = W\,\tfrac{1}{2}l\cos\theta + \tfrac{1}{4}W\{a - b + \sqrt{(a^2 + l^2 - 2al\cos\theta)}\}.$$

Therefore

$$5\frac{d\bar{z}}{d\theta} = -2l\sin\theta + \frac{al\sin\theta}{\sqrt{(a^2 + l^2 - 2al\cos\theta)}},$$

and $$5\frac{d^2\bar{z}}{d\theta^2} = -2l\cos\theta + \frac{al\cos\theta}{\sqrt{(a^2 + l^2 - 2al\cos\theta)}} - \frac{a^2l^2\sin^2\theta}{(a^2 + l^2 - 2al\cos\theta)^{\frac{3}{2}}}.$$

It follows that $d\bar{z}/d\theta$ vanishes for $\theta = 0$, i.e. when AB is vertically upwards, and that for this value of θ

$$5\frac{d^2\bar{z}}{d\theta^2} = -2l + \frac{al}{\sqrt{(a^2 + l^2 - 2al)}}.$$

Also if $l < a$, this expression $= -\dfrac{l(a-2l)}{a-l}$, and is therefore positive if also $a < 2l$, so that $\bar{z}$ is a minimum in the vertical position and it is a stable position of equilibrium.

$d\bar{z}/d\theta$ also vanishes when $\cos\theta = (3a^2+4l^2)/8al$ which gives a real value for θ when $l < a < 2l$, so that there is also a configuration of equilibrium in which the rod is inclined to the vertical.

(iv) *A uniform rectangular heavy plate is suspended from a smooth peg fixed in a vertical wall, by means of a string (length 2a) the extremities of which are attached to two fixed points (distant 2c apart) placed symmetrically on one of the edges of the plate. The dimensions of the plate are known; shew that under certain circumstances there are three positions of equilibrium of which two are stable.* [S.]

Let $HKLM$ be the plate, C the middle point of the side HM to which the string is attached at S and S'. Let P be the peg, then since

$$SP + PS' = 2a,$$

therefore P lies on an ellipse of foci S, S' and major axis of length $2a$.

Since the peg is smooth the tensions in PS, PS' must be equal and therefore PS, PS' are equally inclined to the vertical. But the normal to the ellipse bisects the angle SPS', therefore the normal at P is vertical, and if the normal meets the radius conjugate to CP in F then CF is horizontal.

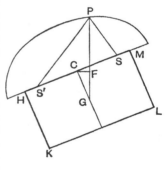

Let G be the centre of the plate and $GC = h$.

For a geometrical solution the possible positions of equilibrium could be found by expressing the fact that the vertical through P passes through G, or that P must be the foot of a normal drawn from G to the ellipse. But in order to discuss the stability we apply the energy test thus:

If the side HM makes an angle θ with the vertical

$$PF^2 = a^2\cos^2\theta + (a^2 - c^2)\sin^2\theta$$

$$= a^2 - c^2\sin^2\theta,$$

and the depth of G below the level of the peg is

$$\bar{z} = h\sin\theta + \sqrt{(a^2 - c^2\sin^2\theta)}.$$

Therefore $\dfrac{d\bar{z}}{d\theta} = h\cos\theta - \dfrac{c^2\sin\theta\cos\theta}{\sqrt{(a^2 - c^2\sin^2\theta)}},$

and $\qquad \dfrac{d^2\bar{z}}{d\theta^2} = -h\sin\theta - \dfrac{c^2\cos 2\theta}{\sqrt{(a^2 - c^2\sin^2\theta)}} - \dfrac{c^4\sin^2\theta\cos^2\theta}{(a^2 - c^2\sin^2\theta)^{\frac{3}{2}}}.$

Hence $d\bar{z}/d\theta$ vanishes for $\theta = \frac{1}{2}\pi$, i.e. the symmetrical position, and also for $\sin\theta = \dfrac{\pm\,ah}{c\sqrt{(c^2+h^2)}}$. If $h < c^2/\sqrt{(a^2-c^2)}$ this gives real values for θ, in which case there are three positions of equilibrium. For $\theta = \frac{1}{2}\pi$, we have $\dfrac{d^2\bar{z}}{d\theta^2} = -h + \dfrac{c^2}{\sqrt{(a^2-c^2)}}$; and, if $h < \dfrac{c^2}{\sqrt{(a^2-c^2)}}$, then $d^2\bar{z}/d\theta^2$ is positive and the depth $\bar{z}$ is a minimum so that the potential energy is a maximum and the equilibrium is unstable, and it is easily verified that with the same conditions $\sin\theta = \dfrac{\pm\,ah}{c\sqrt{(c^2+h^2)}}$ makes $d^2\bar{z}/d\theta^2$ negative so that the two unsymmetrical positions of equilibrium are stable.

If, however, $h > \dfrac{c^2}{\sqrt{(a^2-c^2)}}$, the symmetrical position is the only position of equilibrium and since $d^2\bar{z}/d\theta^2$ is now negative, $\bar{z}$ is a maximum, the potential energy a minimum and the equilibrium stable.

There remains the case in which $h = \dfrac{c^2}{\sqrt{(a^2-c^2)}}$. When this relation holds, then $d\bar{z}/d\theta$ only vanishes for $\theta = \frac{1}{2}\pi$ and we find for the value of the first derivative which does not vanish

$$d^4\bar{z}/d\theta^4 = -3a^2c^2/(a^2-c^2)^{\frac{3}{2}}.$$

This implies that $\bar{z}$ is a maximum and the potential energy a minimum, so that the symmetrical position is the only position of equilibrium and it is stable.

11·65. When a position of equilibrium is given and its stability is to be discussed, it is sufficient to find the change in the potential energy of the system in a slightly displaced position by any direct method. For example.

A uniform rectangular board of weight W is free to turn about one of its edges, and is supported in a horizontal position by a cord which is attached at one end to the middle point of the opposite edge of the board, passes vertically upwards from that point over a smooth fixed pulley of any radius, and carries a weight $\frac{1}{2}W$ at the other end. Prove that, if the board turns through a small angle θ, the potential energy is increased by $Wa^2\theta^4/4h$ approximately, where $2a$ is the breadth of the board, and h the height of the centre of the pulley above the horizontal plane passing through the fixed edge. Hence shew that the horizontal position is one of stable equilibrium.

[T.]

The two figures shew the board AB when horizontal and when turned through an angle θ into the position AB'. In both figures $BC = h$, and in the second figure $B'L$ is at right angles to BC.

The gain in potential energy due to the displacement is

$$Wa\sin\theta - \tfrac{1}{2}W\,(BC - B'C).$$

But $B'C = (LC^2 + B'L^2)^{\frac{1}{2}}$, and $B'L = 2a\,(1-\cos\theta)$,

so that $B'C = \{(h - 2a\sin\theta)^2 + 4a^2\,(1-\cos\theta)^2\}^{\frac{1}{2}};$

and, since we require a result correct to the fourth power of θ, we substitute $\theta - \dfrac{\theta^3}{6}$ for $\sin \theta$, and $\dfrac{\theta^2}{2}$ for $1 - \cos \theta$, giving

$$B'C = h \left\{ 1 - \frac{4a}{h}\theta + \frac{4a^2}{h^2}\theta^2 + \frac{2}{3}\frac{a}{h}\theta^3 - \frac{1}{3}\frac{a^2}{h^2}\theta^4 \right\}^{\frac{1}{2}}$$

$$= h - 2a\theta + \frac{1}{3}a\theta^3 + \frac{1}{2}\frac{a^2}{h}\theta^4 + \text{higher powers of } \theta.$$

Therefore the gain in potential energy

$$= Wa(\theta - \tfrac{1}{6}\theta^3 + ...) - \tfrac{1}{2}W\left(2a\theta - \frac{1}{3}a\theta^3 - \frac{1}{2}\frac{a^2}{h}\theta^4 ...\right)$$

$$= \frac{1}{4}\frac{Wa^2}{h}\theta^4 \text{ approximately.}$$

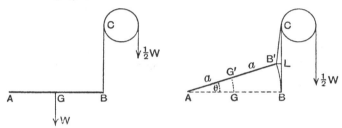

Since this expression is positive for both positive and negative values of θ, it follows that every small displacement causes an increase in potential energy, so that the horizontal position is one of minimum potential energy and stable equilibrium.

11·7. Rocking Cylinders and Spheres.

Let the figure re-present the cross sections of two cylinders whose axes are hori-zontal, one of which of centre of gravity G is free to roll on the rough surface of the other which is fixed.

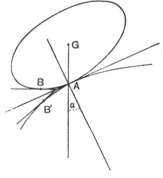

In a position of equilibrium G is vertically above the point of con-tact A. Let $AG = h$ and let the common normal at A make an angle α with the vertical. Consider a small displacement in which the upper curve rolls on the lower bringing the point B on the upper into contact with B' on the lower.

Let the arc $AB = AB' = ds$, and let the tangents at B and B' make angles $d\phi$ and $d\phi'$ with the common tangent at A. Then the angle through which the upper body rolls is $d\phi + d\phi'$, or $ds\left(\dfrac{1}{\rho}+\dfrac{1}{\rho'}\right)$, where ρ, ρ' are the radii of curvature of the curves AB, AB' at A.

Now by the principle of virtual work, to the first order of the small angle turned through, G moves horizontally, and A being the instantaneous centre of rotation*, the horizontal displacement of G is

$$AG\,(d\phi + d\phi'), \quad \text{or} \quad h\,ds\left(\frac{1}{\rho}+\frac{1}{\rho'}\right).$$

If in the displaced position G lies between the verticals at A and B' the weight will have a clockwise moment about the new point of support B' which would tend to restore the upper cylinder to its former position. Hence the equilibrium will be stable if

$$h\,ds\left(\frac{1}{\rho}+\frac{1}{\rho'}\right) < AB'\cos\alpha$$

or if

$$\frac{\cos\alpha}{h} > \frac{1}{\rho}+\frac{1}{\rho'}.$$

Similarly if $\dfrac{\cos\alpha}{h} < \dfrac{1}{\rho}+\dfrac{1}{\rho'}$, the equilibrium will be unstable; and if

$$\frac{\cos\alpha}{h} = \frac{1}{\rho}+\frac{1}{\rho'},$$

the equilibrium is said to be neutral to a first approximation, for to this order the new position of G will lie in the vertical through the new point of contact of the cylinder.

11·8. Hooke's Law. As many problems on work involve the stretching of elastic strings, it will be convenient here to state the rule which connects the tension of such a string with its extension and find an expression for the potential energy stored up in a stretched string†.

The 'extension' of a stretched elastic string means the ratio of the increment in length to the unstretched length. Thus if

* As in *Dynamics*, 5·42.
† See also *Dynamics*, 7·3–7·32.

l_0 is the natural or unstretched length and the stretched length is l, then the extension is $(l - l_0)/l_0$.

Hooke's Law is that the *tension of the string is proportional to the extension*. If T denotes the tension and we state the law in the form

$$T = \frac{\lambda (l - l_0)}{l_0},$$

then λ is called the modulus of elasticity of the string.

To find the work done in stretching a string from its natural length l_0 to a length l.

When the length is x the tension is $\dfrac{\lambda (x - l_0)}{l_0}$ and the work done against the tension in making a small increment dx is

$$\frac{\lambda (x - l_0)}{l_0}\, dx.$$

Therefore the whole work done

$$= \frac{\lambda}{l_0} \int_{l_0}^{l} (x - l_0)\, dx = \frac{1}{2}\frac{\lambda}{l_0} (l - l_0)^2$$
$$= \tfrac{1}{2} T (l - l_0),$$

if T denotes the final tension.

This work is stored up in the string as potential energy.

In like manner the work done in increasing the length from l to l'

$$= \frac{\lambda}{l_0} \int_{l}^{l'} (x - l_0)\, dx$$
$$= \frac{1}{2}\frac{\lambda}{l_0} \{(l' - l_0)^2 - (l - l_0)^2\}$$
$$= \frac{1}{2} (l' - l) \left\{ \frac{\lambda(l' - l_0)}{l_0} + \frac{\lambda(l - l_0)}{l_0} \right\}$$

= the increment in length × the mean of the initial and final tensions.

11·81. Hooke's Law also gives the relation between the tension or thrust and the extension or compression of a spiral spring.

EXAMPLES

1. A smooth circular cylinder of radius b is fixed parallel to a smooth vertical wall with its axis at a distance c from the wall. A smooth uniform heavy rod of length $2a$ rests on the cylinder with one end on the

wall and in a plane perpendicular to the wall, shew that its inclination θ to the horizontal is given by

$$a \cos^3 \theta + b \sin \theta = c.$$ [S.]

2. Two equal light rods AOB, COD, freely jointed at O, their middle point, are at rest in a vertical plane with their ends B, C on a smooth horizontal table. A string to the ends of which equal weights are attached passes over A and D. Shew that in the position of equilibrium the angle between the rods is $\tan^{-1} \frac{4}{3}$.   [I.]

3. A smooth ring is fixed above a smooth plane inclined at an angle α to the horizon. Prove that the rod of least length which passes through the ring and can rest in equilibrium with one end on the plane makes an angle ϕ with the horizontal given by $\sin(\alpha + 2\phi) = 3 \sin \alpha$. [S.]

4. A rectangular lamina $ABCD$ rests with the sides AB, AD on two smooth pegs in a horizontal line; prove that, if the distance between the pegs is half a diagonal of the rectangle, AB, AD bisect the angles between AC and the horizon.   [S.]

5. A uniform square lamina rests in equilibrium in a vertical plane under gravity with two of its sides in contact with smooth pegs in the same horizontal line at a distance c apart. Shew that the angle θ made by a side of the square with the horizontal in a non-symmetrical position of equilibrium is given by

$$c(\sin \theta + \cos \theta) = a,$$

$2a$ being the length of a side of the square.   [S.]

6. Shew by the Principle of Virtual Work that if any number of forces P, Q, R, etc. act on a particle O and ρ is their resultant; and if a transversal be drawn cutting the lines of action of the forces in L, M, N, etc., and the line of action of the resultant in σ,

$$\frac{\rho}{O\sigma} = \Sigma \frac{P}{OL}.$$ [C.]

7. A rod, whose centre of gravity divides it into lengths a and b, is fitted at its ends with rings which can slide on two smooth wires fixed in a vertical plane each at an angle α to the vertical; prove that in equilibrium the depth of the centre of gravity below the intersection of the wires is equal to

$$\tfrac{1}{2} \operatorname{cosec} \alpha \, (a^2 + b^2 + 2ab \cos 2\alpha)^{\frac{1}{2}}.$$ [C.]

8. Four equal uniform rods, of length $a + b$, are hinged at A, B, C, D, and are suspended by two strings of equal length, so that the diagonal AC is vertical and A is the highest point. The strings are attached to two pegs in the same horizontal line and to two points in AB, AD at distance a from A. Prove that if the rods rest in the form of a square the inclination of the strings to the horizontal is $\tan^{-1}(a/b)$.   [S.]

9. A Toggle joint consists of two links, AB, BC, each 1 ft. long, hinged together at B: A is fixed and C can slide without friction along

the fixed line AC. If a force P be applied at B to AB, at right angles to it, determine the force which applied at C will maintain equilibrium when the angle ABC is $120°$. [I.]

10. A pentagon is formed of five equal uniform rods smoothly jointed at their extremities. It hangs with the two upper rods in contact with smooth pegs in the same horizontal line and the lowest rod horizontal. Shew that, if in equilibrium the pentagon is regular, the pegs must divide the rods in the ratio $2 + \sqrt{5} : 3$. [S.]

11. Two equal rods AB, BC each of length l have a smooth joint at B and the ends A, C carry smooth light rings which can slide on the upper arc of an elliptic wire of axes $2a$, $2b$; the major axis $(2a)$ is vertical and the plane of the wire is also vertical. Shew that if $b^2/2a < l < b$, and the system is in equilibrium, each rod is either vertical or inclined to the vertical at an angle θ, where

$$b^4 = l^2 (4a^2 \cos^2 \theta + b^2 \sin^2 \theta).$$ [S.]

12. Two smooth wires OA and OB are fixed in a vertical plane and are inclined at equal angles α to the vertical. Two equal heavy uniform rods HC and KC are smoothly hinged at C and the ends H and K are constrained to slide along OA and OB respectively by means of small smooth weightless rings; the weight of each rod is W, and an additional weight W is attached at C. Shew that in the symmetrical position of equilibrium each rod is inclined at an angle θ to the vertical given by

$$\tan \theta = \tfrac{3}{2} \cot \alpha.$$ [S.]

13. Two particles of masses m and m' are connected by a string of length l resting on a smooth cycloid with its vertex upwards, and base horizontal. Prove that in the equilibrium position the distance of the particle m from the vertex measured along the arc is $m'l/(m + m')$. [C.]

14. Within a smooth fixed elliptic cylinder, whose cross section is an ellipse having its major axis horizontal and equal to twice the vertical minor axis $2b$, is placed a pair of compasses formed of two equal uniform rods, each of length $2b$, jointed together. Shew that any position of the compasses will be one of equilibrium provided that the feet of the compasses meet the cylinder at points on the same level and their plane is perpendicular to the axis of the cylinder. [S.]

15. A rhombus $ABCD$ is formed of four equal uniform rods freely jointed together and suspended from the point A; it is kept in position by a light rod joining the mid-points of BC and CD; prove that if T be the thrust in this rod and W the weight of the rhombus $T = W \tan \tfrac{1}{2} A$. [S.]

16. Four equal uniform rods of weight W are freely jointed so as to form a square $ABCD$ which is suspended from A and is prevented from collapsing by an inextensible string joining the middle points of AB and BC. Prove that the tension of the string is $4W$ and find the magnitude and direction of the reaction at B. [S.]

17. $ABCD$ is a rhombus of freely jointed rods lying flat on a smooth table and P, Q are the middle points of AB, AD. Prove that if the system is held in equilibrium by tight strings joining P to Q and A to C, the tensions in these strings are in the ratio of $2BD$ to AC.   [S.]

18. $ABCD$ is a rhombus formed of light rods loosely jointed together; OB, OD are two equal rods jointed at O, B and D. If O is connected to A and C by strings in tension, prove that the tension of each string is inversely proportional to its length.   [S.]

19. Four equal heavy rods freely jointed at the ends form a square $ABCD$, which when balanced over its lowest point A is kept from changing its form by a light string joining B, D. The mass of each rod being m, find, by the method of virtual work or otherwise, the tension of the string when a weight of mass M is attached to C.   [S.]

20. Four uniform rods are jointed to form a rectangle $ABCD$. AB is fixed in a vertical position with A uppermost, and the rectangle is kept in shape by a string joining AC. Find the tension of the string.   [S.]

21. Three equal rods AB, BC, CD are jointed together smoothly at B and C whilst A and D are fixed at a distance apart equal to the length of either rod. B and C are fastened by strings to points E and F in the same straight line with AD such that $EA = AD = DF$. Shew that the tensions in these strings are proportional to their lengths.   [S.]

22. A framework of n equal weightless rods freely jointed together is maintained in the form of a regular polygon by equal forces P applied at the middle point of each rod at right angles to it. Prove that the tension or thrust in each rod is $\frac{1}{2}P\cot\frac{\pi}{n}$.   [S.]

23. A regular octagon $ABCDEFGH$ is formed of eight equal heavy rods jointed together; the rod AB is fixed horizontally and the framework hangs from it, the regular octagonal form being maintained by two weightless rods joining C and H, D and G. Shew that the reactions along the rods CH and DG are $3W$ and W respectively, where $W =$ weight of one of the heavy rods.   [S.]

24. A regular octahedron formed of twelve equal rods of weight w freely jointed is suspended from one corner. Prove that the thrust in each horizontal rod is $3w/\sqrt{2}$.   [S.]

25. A regular pentagon $ABCDE$ is formed of five uniform heavy rods each of weight W freely jointed at their extremities. It is freely suspended from A and is maintained in its regular pentagonal form by a light rod joining B and E. Prove that the stress in this rod is $W\cot 18°$.   [S.]

26. A triangular lamina ABC rests with its plane vertical, and with the sides AB, AC supported by smooth pegs D, E in a horizontal line. Prove that, if $AD = p$, $AE = q$, then
$$b(q - p\cos A) - c(p - q\cos A) + 3(p^2 - q^2) = 0.   [I.]$$

27. A square $ABCD$ formed of light rods, loosely jointed, has the side AB fixed. The middle points of AB, BC are joined by a string which is kept taut by a force P acting at the middle point of AD parallel to AB. Shew that the tension of the string is equal to $P\sqrt{2}$.

[S.]

28. A rhombus $ABCD$ of loosely jointed rods is in a horizontal plane with the rod BC fixed in position. The middle points of AD, DC are joined by a string which is kept taut by a couple L applied to the rod AB. Prove that the tension of the string is $2L/AB \cos \tfrac{1}{2}ABC$. [I.]

29. A light rhombus formed of rods smoothly jointed at A, B, C, D rests in a vertical plane with A vertically above C and the rods AB, AD over smooth pegs at the same level at a distance $2c$ apart. B, D are connected by a light rod so that the angle A of the rhombus is 2α. Shew that if a weight W is hung from C, the tension in the rod BD is

$W \left(\dfrac{c}{2a} \sec \alpha \operatorname{cosec}^2 \alpha - \tan \alpha \right)$, a being a side of the rhombus. [S.]

30. Three rhombuses $ACBD$, $CEDF$, $EGFH$, each made of four equal light rods loosely jointed, are freely connected at the extremities of their common diagonals CD, EF. The rhombuses are not necessarily in the same plane. Shew that, if the diagonals AB and GH are two stretched strings, their tensions must be proportional to their lengths.

[S.]

31. A rhombus $ABCD$ formed of four uniform freely jointed rods each of weight W and length a rests symmetrically in a vertical plane with AB, AD in contact with two smooth pegs in the same horizontal plane at a distance $2c$ apart (the vertex A being downwards), and is kept from collapsing by a light string BD. Prove that the tension of the string is $2W (a \sin \tfrac{1}{2}A - c \operatorname{cosec}^2 \tfrac{1}{2}A)/a \cos \tfrac{1}{2}A.$ [S.]

32. A parallelogram $ABCD$ is formed of uniform heavy rods freely jointed at their extremities. AB is held fixed in a horizontal position and the parallelogram is maintained in its form so that ADC is an acute angle α by means of a string joining A to a point P in DC. Prove that the tension of the string is $W . AP \cot \alpha / DP$, where W is half the weight of the parallelogram. [S.]

33. A string of length a forms the shorter diagonal of a rhombus formed of four uniform rods, each of length b and weight W, hinged together; prove that, if one of the rods is supported in a horizontal position, the tension of the string is

$$2Wb^{-1}(2b^2 - a^2)(4b^2 - a^2)^{-\frac{1}{2}}.$$ [S.]

34. AB is the horizontal diameter of a circular wire of radius a whose plane is vertical. A bead of mass M at the lowest point C can slide on the wire and is attached to two strings which pass through small fixed rings at A, B. To the other ends of the strings are attached equal masses m which hang freely. Find the potential energy of the

system when it is displaced so that the radius to the bead makes an angle θ with the vertical.

Shew that the equilibrium with M at C is stable if $m < M\sqrt{2}$. [S.]

35. An isosceles triangle of angle 2α rests between two smooth pegs at the same level, distant $2c$ apart; prove that if h is the distance of the centre of gravity from the vertex, and if

$$2c \sec \alpha < h < 2c/\sin \alpha \cos \alpha,$$

then oblique positions of equilibrium exist, which are unstable.

Discuss the stability of the vertical position in case

$$h = 2c/\sin \alpha \cos \alpha. \qquad [\text{C.}]$$

36. A uniform smooth rod passes through a ring at the focus of a fixed parabola, whose axis is vertical and vertex below the focus, and rests with one end on the parabola. Prove that the rod will be in equilibrium if it makes with the vertical an angle θ given by the equation

$$\cos^4 \frac{\theta}{2} = \frac{a}{2c},$$

where $4a$ is the latus rectum and $2c$ the length of the rod.

Investigate also the stability of the equilibrium in this position. [S.]

37. Three equal uniform rods, DA, DB, DC, each of length $2l$ and of weight W, are smoothly jointed together at D, and respectively pass through three small smooth fixed rings at the corners of an equilateral triangle, whose length of side is a and plane horizontal. Also a weight w is attached to D. Prove that a symmetrical position of equilibrium is possible, if

$$\frac{1}{3}\frac{w}{W} < \frac{\sqrt{3}l}{a} - 1.$$

Also prove that two symmetrical positions of equilibrium will then exist, and investigate their stability. [S.]

38. A uniform rod AB of weight W can turn freely round one end A: a fine cord is attached to a point C vertically above A, passes through an eyelet fixed on the rod at the end B and carries a hanging weight w at its other end. Prove that, in the absence of friction, the rod will be in equilibrium when BC is equal to $w.AC/(\frac{1}{2}W+w)$. Prove also that this position of equilibrium is an unstable one. [S.]

39. AB is a uniform rod, of length $6a$ and weight W, which can turn freely about a fixed point in its length distant $2a$ from A. AC and BC are light strings of length $5a$ attached to a particle C of weight w. Shew that if W is less than $2w$ there will be stable equilibrium with AB inclined to the horizontal at an angle $\tan^{-1}\dfrac{W+w}{4w}$. [S.]

40. A uniform heavy rod of length $2l$ rests with its ends on a fixed smooth parabola with axis vertical and vertex downwards (latus rectum $= 4a$). Shew that if $l > 2a$ there are three positions of equilibrium and that the horizontal position is then unstable, but that if $l \leqslant 2a$ the only position of equilibrium is horizontal. [S.]

41. Shew that a rough uniform plank resting horizontally on the top of a circular cylinder will be in stable equilibrium if its thickness is less than the diameter of the cylinder.

Supposing the condition satisfied, find the greatest displacement for which the stability obtains. [C.]

42. A solid sphere rests inside a fixed rough hemispherical bowl of twice its radius. Shew that, however large a weight is attached to the highest point of the sphere, the equilibrium is stable. [S.]

43. A solid circular cylinder of radius a and height h has one end in the shape of a hemisphere; find the condition that it will be in stable equilibrium when standing on that end, on a smooth horizontal plane, with its axis vertical. [S.]

44. A uniform hemisphere rests in equilibrium with its base upwards on top of a sphere of double its radius. Shew that the greatest weight which can be placed at the centre of the plane face without rendering the equilibrium unstable is one-eighth of the weight of the hemisphere. [S.]

45. A homogeneous hemispherical shell of radius a and weight W has a weight $2W/3$ fastened to a point in the rim of the hemisphere. Prove that if properly placed it can rest in neutral equilibrium on the top of a fixed sphere of radius a. [S.]

46. A stiff wire in the form of a parabola rests on the ground with its plane vertical. The centre of gravity of the wire is on the axis of the parabola at a distance h from the vertex, and the latus rectum is $4a$. Prove that, if $h > 2a$, there is a position of equilibrium in which the axis makes an angle $\tan^{-1}\left(\dfrac{a}{h-2a}\right)^{\frac{1}{2}}$ with the horizon. Also prove that this position of equilibrium is stable. [S.]

47. Two equal particles repel each other according to the fifth power of the distance, and are connected by an elastic string. Find the position of equilibrium, and shew that it is stable if the extension of the string is less than one-quarter of its original length. [S.]

48. A uniform elastic string has a length a_1 when the tension is T_1, and a length a_2 when the tension is T_2. Shew that its natural length is

$$\frac{a_2 T_1 - a_1 T_2}{T_1 - T_2},$$

and that the amount of work done in stretching it from its natural length to a length $(a_1 + a_2)$ is

$$\tfrac{1}{2} \cdot \frac{(a_1 T_1 - a_2 T_2)^2}{(T_1 - T_2)(a_1 - a_2)}. \qquad [\text{C.}]$$

49. A uniform rod of weight W and length l is suspended from a fixed point by two light elastic strings attached to its ends. If the strings have the same modulus of elasticity, W, and are of natural

lengths l and $l/2$, prove that their lengths in the position of equilibrium are l/x and l/y, where
$$y = 1 + x,$$
and
$$\frac{1}{x^2} + \frac{1}{1+x^2} = \frac{1}{2}\left\{1 + \frac{1}{(1-x)^2}\right\}. \qquad \text{[S.]}$$

50. OA is a slightly compressible vertical rod of height h and negligible mass (modulus of compressibility μ) freely pivoted at its lowest point O. AB is a slightly extensible cord of natural length l (modulus λ). B is a point in the horizontal plane through O distant a from O where $a^2 = l^2 - h^2$. A horizontal force P is applied at A in the direction BO. Shew that the horizontal and vertical components of the displacement of A are approximately (neglecting x^2 and y^2)
$$x = \frac{P}{a^2}\left(\frac{h^3}{\mu} + \frac{l^3}{\lambda}\right), \quad y = \frac{Ph^2}{a\mu}. \qquad \text{[S.]}$$

51. Two small heavy rings connected by a light elastic string can slide without friction one on each of two fixed straight wires OA, OB, which lie in a vertical plane through O, the highest point, and are both inclined to the vertical at $45°$. Prove that there is only one configuration of equilibrium, and that if the weights of the rings are $\frac{1}{8}$ and $\frac{7}{8}$ of the modulus of elasticity of the string, the length of the string is twice its natural length.

Investigate the stability of this configuration. [S.]

52. The sides of a parallelogram $ABCD$ are four stretched extensible strings with their ends tied to the two straight rods AC and BD which form the diagonals. The natural lengths of AD and BC are a, and those of AB and DC are b; the stretched length BC is r. Prove that, if the system is in equilibrium, r satisfies the equation
$$(c^2 + d^2 - 2r^2)\left(\frac{1}{b} - \frac{1}{a} + \frac{1}{r}\right)^2 = 2,$$
where c and d are the lengths of AC and BD, and all the strings have the same modulus of elasticity. [S.]

53. The lower ends of three identical vertical springs of length l_0 and large modulus of elasticity λ are attached to three fixed points at equal distances a apart in a horizontal line. A bar of mass M is placed across their upper ends and attached to them in a position in which its centre of gravity is at a distance c from the middle spring. Find the potential energy when the middle spring is compressed a distance x, and the rod makes a small angle θ with the horizontal, and hence find the position of equilibrium and for what values of c/a one of the end springs becomes extended. [S.]

54. Two equal uniform rods are hinged together at one of their ends, and the other ends are connected by a light elastic string whose natural length is equal to the length of either rod. They are placed with the hinge upwards, in a vertical plane, and resting on a smooth horizontal plane. If it be assumed that the tension of an elastic string varies as

the amount of stretching, and if a tension equal to the weight of either
rod would stretch the string to double its natural length, shew that in
the position of equilibrium the inclination of either rod to the horizon
is given by

$$4 = \frac{1}{\sin\theta} + \frac{2}{\cos\theta}.$$ [S.]

ANSWERS

9. P. 16. W vertically, $\frac{1}{2}W$ horizontally.

19. $(M+2m)g$.

20. $W.AC/2AB$, where W = weight of rectangle.

34. $2\sqrt{2}mga\cos\frac{1}{2}\theta - Mga\cos\theta + \text{const.}$

35. Unstable. 36. Stable.

37. Stable or unstable according as D is above or below the triangle.

41. Stable so long as $\cos^2\theta >$ (thickness)/(diameter).

43. $h < a(1+1/\sqrt{2})$. 51. Stable.

53. $Mg(l_0 - x - c\theta) + \frac{1}{2}\lambda(3x^2 + 2a^2\theta^2)/l_0$; $x = \frac{1}{3}Mgl_0/\lambda$, $\theta = \frac{1}{2}Mgl_0c/\lambda a^2$;
$c/a > \frac{2}{3}$.

Chapter XII

FLEXIBLE CHAINS AND STRINGS

12·1. Equations of equilibrium of a chain in one plane under the action of any given forces. Let s be the length of the chain measured from some fixed point A up to a variable point P, and δs the length of the small arc PQ. When the chain is acted upon by external forces the tension will not be constant. It is convenient to assume that the tension in-

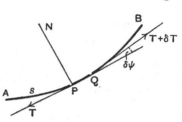

creases in the sense in which we take the arc s to increase, so let T denote the tension at P and $T + \delta T$ denote the tension at Q. Also let $\delta\psi$ denote the small angle between the tangents at P and Q.

Now when we consider the equilibrium of the element PQ of the chain we have to take into account not only the external forces, such as its weight, which act upon it, but also the pull of the rest of the chain upon it; and this pull is represented by a tension T along the tangent at P in the sense PA and a tension $T + \delta T$ along the tangent at Q in the sense QB.

Let us resolve these two forces along the tangent at P in the sense in which s increases, and along the inward normal to the curve AB at P. We get

$$-T + (T + \delta T)\cos\delta\psi \text{ along the tangent}$$

and $\quad (T + \delta T)\sin\delta\psi$ along the inward normal.

To the first order of small quantities these expressions reduce to $\qquad \delta T$ along the tangent ....(1),

and $\qquad T\delta\psi$ along the inward normal ....(2).

These expressions are of great importance; they constitute a convenient measurement of the reaction of the rest of a chain upon an element of itself; viz. that this reaction is compounded of a force δT along the tangent to the element and a force $T\delta\psi$ along the inward normal.

In order to write down equations of equilibrium, our only further requirement is a specification of the external force system. The external forces acting upon an element PQ of the chain will necessarily be proportional to the length δs of the element, and may therefore be represented by a force $F\delta s$ along the tangent at P (in the sense PQ) and a force $G\delta s$ along the inward normal at P; i.e. F and G denote components of force, per unit length of chain.

The equations of equilibrium of the element PQ are then

$$\delta T + F\delta s = 0 \quad \dots\dots\dots\dots\dots\dots(3),$$

and
$$T\delta\psi + G\delta s = 0 \quad \dots\dots\dots\dots\dots\dots(4),$$

which may also be written

$$\frac{dT}{ds} + F = 0 \quad \dots\dots\dots\dots\dots\dots(5),$$

and
$$\frac{T}{\rho} + G = 0 \quad \dots\dots\dots\dots\dots\dots(6),$$

where ρ is the radius of curvature of the curve formed by the chain.

12·11. Special cases. In what follows we shall suppose that the chain or string and the forces acting upon it lie in the same plane unless the contrary is expressly stated.

(i) *A string in contact with a smooth surface under the action of no forces but the reaction of the surface and tensions applied at its ends.*

Here there is no tangential component of external force, so that equation (3) is simply

$$\delta T = 0$$

or
$$T = \text{constant},$$

along the whole length of the string.

Again, if R denotes the reaction of the surface per unit length of string along the *outward* normal, equation (4) gives

$$T\,\delta\psi - R\,\delta s = 0$$

or

$$R = \frac{T}{\rho},$$

where ρ is the radius of curvature; so that T, being constant, the reaction at any point varies as the curvature.

12·12. (ii) *A string in limiting equilibrium in contact with a rough surface and under the action of no forces but the reaction of the surface and tensions applied at its ends.*

Let AB be the string just about to slip from A towards B under the action of tensions T_A, T_B at its ends.

Then with the notation of **12·1**, let the tangents at A, P, B make angles α, ψ, β with a fixed direction.

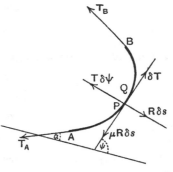

The reaction of the surface on an element PQ of length δs may be represented by a force $R\,\delta s$ along the outward normal and a force $\mu R\,\delta s$ along the tangent opposed to the direction of slipping. Then equations (3) and (4) of **12·1** become

$$\delta T - \mu R\,\delta s = 0$$

and

$$T\,\delta\psi - R\,\delta s = 0.$$

Whence we get

$$\frac{\delta T}{T} = \mu\,\delta\psi,$$

or

$$\log T = \mu\psi + \text{constant},$$

or

$$T = Ce^{\mu\psi}.$$

But when $\psi = \alpha$ then $T = T_A$, therefore

$$T_A = Ce^{\mu\alpha},$$

and, by division

$$T = T_A\,e^{\mu(\psi-\alpha)} \quad\ldots\ldots\ldots\ldots\ldots\ldots(1).$$

In particular

$$T_B = T_A\,e^{\mu(\beta-\alpha)} \quad\ldots\ldots\ldots\ldots\ldots(2).$$

It follows that the tension increases along the string in the sense in which it is about to slip by this exponential factor, in which the angle is the angle turned through by the tangent to the string as we proceed along it.

12·121. The result of **12·12** is of great practical importance and explains for example how it is possible for a single man, by hitching a rope round a post, to destroy the momentum of a vessel of some size arriving at a landing stage.

Example. *Find the ratio of the tensions when a rope is coiled three times round a post and just about to slip, taking* $\mu = 0.5$.

The required ratio $= e^{.5 \times 6\pi} = e^{3\pi}$.

But $\log_{10} e^{3\pi} = 3\pi \times .4343$
$= 4.093$;

therefore the ratio is as $12,400 : 1$.

12·13. (iii) *A heavy uniform string in contact with a smooth curve in a vertical plane.*

Let PQ be an element of the string AB. The string may either be regarded as resting on the upper side of the curve, or as pressed against the lower side of the curve by applying tensions

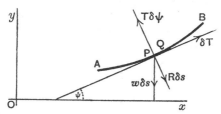

at its ends; and the normal reaction $R\delta s$ of the curve on an element δs of the string will be inwards or outwards in the two cases. In the figure the latter case is taken.

Let ψ denote the angle which the tangent at P makes with a horizontal axis Ox, and let P be the point (x, y). The arc AP is measured upwards so that s and ψ increase together. Let w denote the weight of unit length of the string.

Then by resolving along the tangent and normal at P we get
$$\delta T - w\,\delta s \sin\psi = 0,$$
and $$T\,\delta\psi - w\,\delta s \cos\psi - R\,\delta s = 0,$$
or since $\sin\psi = dy/ds$, $T = wy + \text{const.}$ ....(1),

and $$R = \frac{T}{\rho} - w\cos\psi \quad ....(2).$$

Equation (1) determines the tension at any point when its value is known at some one point and equation (2) then determines the pressure at any point.

It follows from (1) that, if T_1, T_2 be the tensions at points whose ordinates are y_1, y_2, then

$$T_1 - T_2 = w(y_1 - y_2),$$

or the difference between the tensions at two points is proportional to the difference between their levels.

12·131. Example. *A heavy uniform string passes round a circular cylinder whose axis is horizontal. Find the tension of the string at the points where it is vertical, if the pressure on the cylinder vanishes at the lowest point.*

Let a be the radius of the cylinder, w the weight of unit length of the string.

Consider a small element of length $a\,\delta\theta$, where the radius makes an angle θ with the downward vertical.

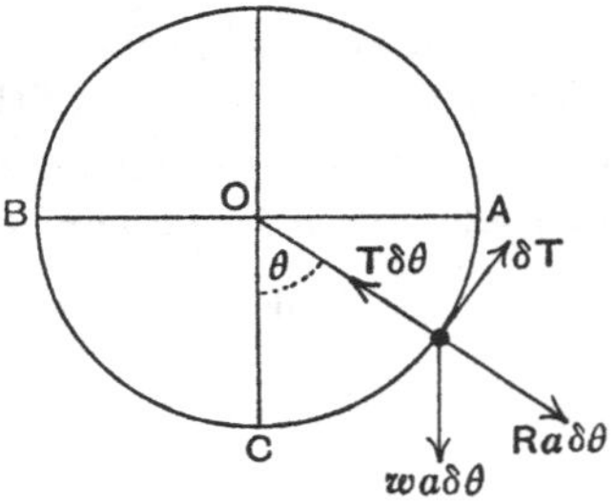

The forces acting on this element are as shewn in the figure, and, by resolving along the tangent and normal, we get

$$\delta T = wa\,\delta\theta \sin\theta,$$

or $\qquad T = C - wa\cos\theta \quad......(1);$

and $\qquad T\,\delta\theta = Ra\,\delta\theta + wa\,\delta\theta\cos\theta,$

or $\qquad T = Ra + wa\cos\theta \quad ....(2).$

If R vanishes when $\theta = 0$, the tension at the lowest point is from (2) $T = wa$, and substituting this value for T in (1) when $\theta = 0$, gives $C = 2wa$.

Therefore $\qquad T = 2wa - wa\cos\theta,$

and where the string is vertical $\theta = \tfrac{1}{2}\pi$, so that the required tension is $2wa$.

12·14. (iv) *A heavy uniform string in limiting equilibrium in contact with a rough curve in a vertical plane.*

Using the notation and figure of **12·13**, if the string is about to slip from A towards B, there is an additional tangential force on the element PQ of amount $\mu R\,\delta s$, due to friction, and, by resolving along the tangent and normal at P we get

$$\delta T - \mu R\,\delta s - w\,\delta s \sin\psi = 0,$$

and $\qquad T\,\delta\psi - R\,\delta s - w\,\delta s \cos\psi = 0.$

Eliminating R and dividing by $\delta\psi$ we get

$$\frac{dT}{d\psi} - \mu T = w\rho\,(\sin\psi - \mu\cos\psi) \quad\ldots\ldots\ldots\ldots(1),$$

where ρ is the radius of curvature of the curve at P.

This equation is integrated by multiplying both sides by the integrating factor $e^{-\mu\psi}$, and leads to

$$Te^{-\mu\psi} = C + w\int e^{-\mu\psi}\rho\,(\sin\psi - \mu\cos\psi)\,d\psi \quad\ldots\ldots(2).$$

For a curve of given intrinsic equation, ρ can be expressed in terms of ψ, and in some cases the integral in (2) can be found by the aid of the standard forms

$$\int e^{-\mu\psi}\sin a\psi\,d\psi = \frac{-e^{-\mu\psi}}{a^2+\mu^2}(a\cos a\psi + \mu\sin a\psi),$$

$$\int e^{-\mu\psi}\cos a\psi\,d\psi = \frac{e^{-\mu\psi}}{a^2+\mu^2}(a\sin a\psi - \mu\cos a\psi).$$

12·141. Example. *A heavy string occupies a quadrant of the upper half of a rough vertical circle in a state bordering on motion. Prove that the radius through the lower end makes an angle α with the vertical given by* $\tan(\alpha - 2\lambda) = e^{-\frac{1}{2}\pi\mu}$, *where* $\mu = \tan\lambda$ *is the coefficient of friction.*

Let AB be the string, a the radius of the circle, PQ a small element of the string of length $a\,\delta\theta$, where θ is the angle which the radius OP makes with the vertical. Then using T, R and w as in **12·14**, the forces acting on the element PQ are as shewn in the figure, and the equations of equilibrium are

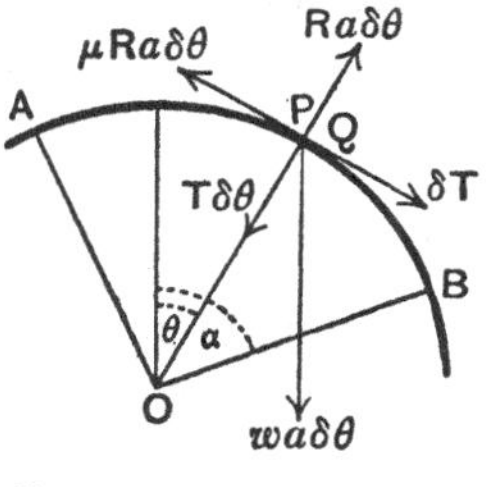

$$\delta T - \mu Ra\,\delta\theta = -wa\,\delta\theta\sin\theta,$$

and $\qquad T\,\delta\theta - Ra\,\delta\theta = -wa\,\delta\theta\cos\theta,$

leading to the differential equation

$$\frac{dT}{d\theta} - \mu T = -wa\,(\sin\theta - \mu\cos\theta).$$

This gives on integration

$$Te^{-\mu\theta} = C - \frac{wae^{-\mu\theta}}{1+\mu^2}\{(\mu^2-1)\cos\theta - 2\mu\sin\theta\},$$

or putting $\tan\lambda$ for μ

$$T = Ce^{\mu\theta} + wa\cos(\theta - 2\lambda).$$

Now T vanishes at both ends of the string, i.e. at A where

$$\theta = -(\tfrac{1}{2}\pi - \alpha),$$

and at B where $\qquad\qquad \theta = \alpha.$

Therefore $\qquad 0 = Ce^{-\mu(\frac{1}{2}\pi-\alpha)} + wa\sin(\alpha - 2\lambda),$

and $\qquad\qquad 0 = Ce^{\mu\alpha} + wa\cos(\alpha - 2\lambda);$

so that, by division, $\qquad \tan(\alpha - 2\lambda) = e^{-\frac{1}{2}\pi\mu}.$

12·2. The Common Catenary. This is the curve in which a uniform chain or a 'perfectly flexible' string hangs when freely suspended from two fixed points.

At the lowest point C of the curve the tension is horizontal and denoted by T_0.

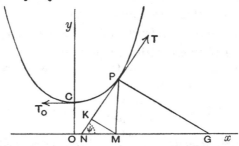

Let s be the length of the arc measured from C to any point P of the chain. Let T be the tension at P and ψ its inclination to the horizontal.

If w denotes the weight of a unit of length of the chain, the weight of the portion CP is ws, and the portion CP is in equilibrium under the action of three forces, viz. its weight and the tensions T_0, T at its ends.

Therefore, by resolving horizontally and vertically, we get

$$T \cos \psi = T_0 \quad\text{.......................(1)},$$

and

$$T \sin \psi = ws \quad\text{.......................(2)}.$$

It is convenient to introduce another constant and write $T_0 = wc$ in (1); then by division we obtain the equation

$$s = c \tan \psi \quad\text{.......................(3)},$$

which is the intrinsic equation of the catenary.

Since c is the only constant in the equation, it is called *the parameter of the catenary*.

The Cartesian equation is easily deduced thus:

By differentiating $\quad s = c \tan \psi = c \dfrac{dy}{dx}$,

we obtain $\quad\quad\quad\quad \dfrac{ds}{dx} = c \dfrac{d^2 y}{dx^2}$,

or $\quad\quad\quad\quad \sqrt{\left\{ 1 + \left(\dfrac{dy}{dx} \right)^2 \right\}} = c \dfrac{d^2 y}{dx^2}$.

This gives on integration

$$c \sinh^{-1} \frac{dy}{dx} = x + A,$$

where A is a constant of integration.

The positions of the axes of co-ordinates are not yet fixed. Let us take the vertical through the lowest point of the chain for axis of y; then $dy/dx = 0$ when $x = 0$, so that $A = 0$ and

$$\frac{dy}{dx} = \sinh \frac{x}{c} \quad \dotfill (4).$$

Integrating again we get

$$y = c \cosh \frac{x}{c} + B,$$

where B is a constant of integration.

Now take the origin at a depth c below the lowest point of the chain, and we have $y = c$ when $x = 0$, so that $B = 0$, and

$$y = c \cosh \frac{x}{c} \quad \dotfill (5)$$

is the equation of the catenary; and the axis Ox is called its directrix. The curve is clearly symmetrical about its lowest point.

12·21. Geometrical Properties of the Common Catenary.

Since
$$s = c \tan \psi = c \frac{dy}{dx},$$

and from (4)
$$\frac{dy}{dx} = \sinh \frac{x}{c},$$

therefore
$$s = c \sinh \frac{x}{c} \quad \dotfill (6).$$

Again if PM or y be the ordinate drawn to the directrix and MK be the perpendicular from M to the tangent at P, we have, from (5)

$$y^2 = c^2 \cosh^2 \frac{x}{c} = c^2 + c^2 \sinh^2 \frac{x}{c},$$

therefore, from (6)
$$y^2 = c^2 + s^2 \quad \dotfill (7).$$

But $s = c \tan \psi$, therefore
$$y = c \sec \psi \quad \dotfill (8).$$

Again, in the triangle PKM,

$$MK = MP \cos \psi$$
$$= y \cos \psi$$
$$= c,$$

and
$$PK = KM \tan \psi = c \tan \psi = s;$$

i.e. PK is equal to the arc PC of the curve.

It follows that if a string were wrapt round a material curve in the form of a catenary and then cut at the vertex C and gradually unwrapt, the end of the string would trace out the locus of K in the figure; in other words the locus of K is an involute of the catenary. It is called the tractrix. KP is the normal to the involute, and KM is the tangent, and we have proved that KM, the portion of the tangent to the involute cut off by the directrix, is constant and equal in length to the parameter c of the catenary.

Further, if the normal at P to the catenary meets the directrix in G, we have

$$PG = PM \sec \psi = c \sec^2 \psi$$

$$= \frac{ds}{d\psi} = \rho;$$

so that the radius of curvature at any point on the catenary is numerically equal to the length of the normal intercepted between the curve and the directrix, but they are drawn in opposite directions.

12·22. The Tension. It follows from the Cartesian equation that all common catenaries are similar and that their relative magnitudes depend only on the parameter c.

The tension T at any point is given by **12·2** (1); thus the horizontal component is

$$T \cos \psi = T_0 = wc,$$

so that
$$T = wc \sec \psi = wy \dots\dots\dots\dots\dots(9),$$

i.e. the tension at any point is proportional to the height of the point above the directrix.

It follows that if a chain hangs in festoons over a number of smooth pegs that the catenaries in which it hangs all have the same directrix and that if the chain has free ends they must be on the directrix.

12·23. The Parameter. When a given length of uniform chain hangs between two given points it is a definite problem to find the parameter of the catenary.

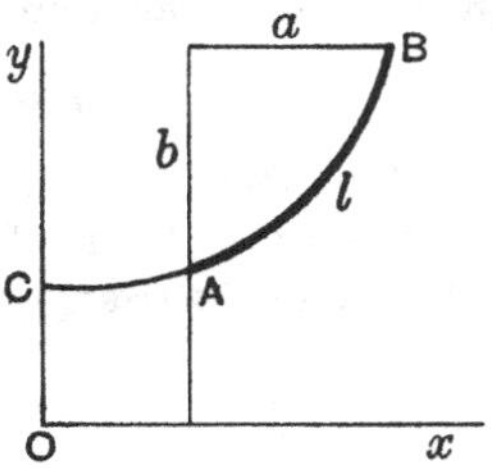

Thus if AB is the chain of length l and a, b are the co-ordinates of B referred to A, and A is the point (x, y) on a catenary of parameter c whose arc from A to B is of length l, we have

$$y = c \cosh \frac{x}{c}, \qquad s = c \sinh \frac{x}{c},$$

and
$$y + b = c \cosh \frac{x+a}{c}, \quad s + l = c \sinh \frac{x+a}{c};$$

so that
$$b = c\left(\cosh\frac{x+a}{c} - \cosh\frac{x}{c}\right),$$

and
$$l = c\left(\sinh\frac{x+a}{c} - \sinh\frac{x}{c}\right).$$

We find that x can be eliminated by squaring and subtracting, and that
$$2c\sinh\frac{a}{2c} = \pm\sqrt{(l^2 - b^2)}.$$

This equation cannot be solved explicitly, but it can be shewn that there is only one positive value of c that will satisfy the equation*, so that only one position of equilibrium is possible. An approximate solution may be found by the help of tables of hyperbolic functions in particular cases.

If the chain is nearly taut, the parameter c is large compared to the length of the string. For if ψ, ψ' are the inclinations to the horizontal at the ends A, B, since $CA = c\tan\psi$ and $CB = c\tan\psi'$, therefore
$$l = c(\tan\psi' - \tan\psi);$$
but ψ and ψ' are nearly equal in this case, so that c must be large compared to l.

12·231. Example. *A uniform chain of length l when tightly stretched between two points at the same level has a sag k in the middle. Prove that the length of the chain exceeds the distance between the points by $8k^2/3l$ approximately.*

If $2a$ be the distance between the points and c the parameter of the catenary, we have

$$k + c = c\cosh\frac{a}{c} \quad\dots\dots\dots\dots\dots\dots(1),$$

and
$$\tfrac{1}{2}l = c\sinh\frac{a}{c} \quad\dots\dots\dots\dots\dots\dots(2),$$

so that
$$(k+c)^2 = c^2 + \tfrac{1}{4}l^2, \text{ and } 2kc = \tfrac{1}{4}l^2 - k^2\dots\dots\dots\dots(3);$$
this gives the parameter of the catenary in the form
$$c = \frac{l^2}{8k} - \frac{k}{2}, \quad\dots\dots\dots\dots\dots\dots(4),$$

where, since k is small, the second term may be neglected in comparison with the first.

Since a/c is small we may expand $\sinh\frac{a}{c}$ in (2) and retain only the first two terms, so that
$$\tfrac{1}{2}l = c\left(\frac{a}{c} + \frac{a^3}{6c^3}\right),$$

or
$$l - 2a = \frac{a^3}{3c^2}.$$

* v. Routh's *Analytical Statics*, vol. I, § 447.

A first approximation $a = \frac{1}{2}l$ may be substituted in the term $a^3/3c^2$, so that

$$l - 2a = \frac{l^3}{24c^2},$$

and substituting for c from (4) we get

$$l - 2a = 8k^2/3l.$$

12·3. The Parabolic Chain. *If particles of equal weight are so distributed along a light string hanging between two fixed points that the horizontal distance between each pair of consecutive particles is the same, then the particles lie on a parabola whose axis is vertical.*

Let O, P, Q, R be four of the particles. Take O as origin and horizontal and vertical axes Ox, Oy.

Let a be the horizontal distance between each pair of consecutive particles, and let y_1, y_2, y_3 denote the ordinates of P, Q, R and θ_1, θ_2, θ_3 the inclinations of OP, PQ, QR to the vertical. Then, from **6·5**, since the weights are equal,

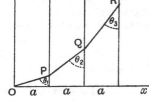

$$\cot \theta_3 - \cot \theta_2 = \cot \theta_2 - \cot \theta_1,$$

or $\dfrac{y_3 - y_2}{a} - \dfrac{y_2 - y_1}{a} = \dfrac{y_2 - y_1}{a} - \dfrac{y_1}{a}.$

Therefore $\qquad\qquad y_3 = 3(y_2 - y_1)$ ....(1).

Now one parabola with axis vertical can be drawn to pass through three assigned points, say O, P, Q, for the equation

$$y = Ax^2 + Bx \qquad\qquad ....(2)$$

represents a parabola with its axis vertical passing through O, and A and B can be so chosen that the curve will pass through any two assigned points.

The co-ordinates of P, Q and R are a, y_1; $2a$, y_2 and $3a$, y_3, and the curve (2) passes through P and Q if

$$y_1 = Aa^2 + Ba,$$
and $\qquad\qquad y_2 = 4Aa^2 + 2Ba.$

Therefore, from (1),

$$y_3 = 3(y_2 - y_1) = 9Aa^2 + 3Ba,$$

shewing that the same parabola passes through R, and similarly through all the points on the chain.

12·31. If the distances between the particles be indefinitely diminished, we arrive at the case of a light string or chain supporting a continuous load whose horizontal distribution is uniform; i.e. the case

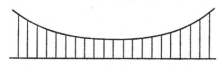

of a suspension bridge in which the weights of other parts are negligible in comparison with that of the horizontal roadway.

The vertical rods that link the roadway to the parabolic chain are spaced at equal horizontal distances and so carry equal loads.

12·32. The parabolic form of a chain carrying a continuous load uniformly distributed *horizontally* may be proved simply as follows:

Let O be the lowest point of the chain, and P any point (x, y) referred to horizontal and vertical axes through O. The weight carried by OP is proportional to ON and may be denoted by wx, and this force acts through the middle point Q of ON, since the load is uniformly distributed horizontally. The portion OP of the chain is also acted upon by the tensions at its ends, say T at P and T_0 at O. The former depends on the position of P on the chain, and the latter is a definite

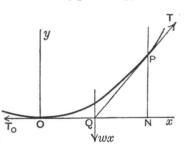

constant. The three forces acting upon the portion OP of the chain must meet in the point Q, and PNQ is a triangle of forces, so that

$$\frac{wx}{PN} = \frac{T_0}{NQ}, \quad \text{or} \quad T_0 y = \tfrac{1}{2} wx^2;$$

which represents a parabola of latus rectum $2T_0/w$.

12·33. Examples. (i) *In a suspension bridge of* 400 *ft. span and* 40 *ft. dip the whole weight supported by the two chains is* 2 *tons per horizontal foot. Find the horizontal tension in the chains and the tension at the point of support.* [T.]

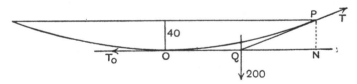

There are two parallel chains each carrying half the load. If OP represents half of one of the chains, the load that it carries is 200 tons, and with the notation of **12·32**

$$QN = 100, \quad \text{and} \quad PN = 40 \quad \text{so that} \quad PQ = 10\sqrt{116} = 107 \cdot 7.$$

Then by the triangle of forces

$$\frac{T_0}{NQ} = \frac{T}{QP} = \frac{200}{PN},$$

or

$$\frac{T_0}{100} = \frac{T}{107 \cdot 7} = \frac{200}{40} = 5,$$

so that

$$T_0 = 500 \text{ tons} \quad \text{and} \quad T = 538 \cdot 5 \text{ tons}.$$

(ii) *If a telegraph wire has a span of* 75 *yd. and a sag in the middle of* 1 *ft., shew that the tension in the wire is approximately* 480 *lb. weight, when the weight of the wire is* 400 *lb. per mile.*   [T.]

Considering the equilibrium of one-half of the wire, its weight is $\frac{75}{2} \times \frac{400}{1760}$ lb. acting at a distance from the end P approximately equal to one-quarter of the span, i.e. $\frac{225}{4}$ ft.

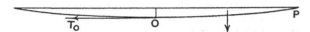

Hence, if T_0 is the tension at the lowest point O, by taking moments about P for the forces acting on the portion OP, we get

$$T_0 = \tfrac{75}{2} \times \tfrac{400}{1760} \times \tfrac{225}{4}$$
$$= 479 \cdot 4 \, \text{lb.}$$

The tension in the wire at other points has a vertical component which is negligible in comparison with the horizontal component because the wire is nearly horizontal at all points. Hence the tension at all points is approximately 480 lb. weight.

12·4. Chain of Variable Density or Thickness.

Let ρ be the density and ω the area of the cross section at any point, then the mass of a small element of length δs at this point is $\rho \omega \, \delta s$, or $m \, \delta s$, where m is variable on account of the variability of ρ or of ω or of both ρ and ω.

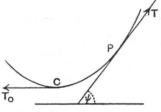

Hence if T_0 and T denote the tensions at the lowest point C and at a point P, where $CP = s$ and the tangent at P makes an angle ψ with the horizontal, by resolving horizontally and vertically for the portion CP, we get

$$T \cos \psi = T_0$$

and
$$T \sin \psi = \int_0^s mg \, ds;$$

therefore
$$T_0 \tan \psi = \int_0^s mg \, ds,$$

and, by differentiation,

$$T_0 \sec^2 \psi \frac{d\psi}{ds} = mg,$$

or
$$m \cos^2 \psi \frac{ds}{d\psi} = \text{constant} \dots\dots\dots\dots(1).$$

If the form of the curve be given, this equation determines the value of m; and, if m be given, the integral of equation (1) gives the intrinsic equation of the curve.

12·41. Example. *Find the law of density of a chain which hangs in a cycloid.*

The intrinsic equation of a cycloid is
$$s = 4a \sin \psi,$$
so that **12·4** (1) gives
$$m \cos^3 \psi = \text{constant},$$
or
$$m \propto \sec^3 \psi.$$

12·5. Catenary of Uniform Strength. I.e. a chain of uniform material, suspended from two fixed points, and such that the area of its cross section is proportional to the tension.

It follows from this definition that the weight of an element is proportional to the cross section so that if $w \, \delta s$ denotes the weight of an element δs at a place where the tension is T, then $T = wc$, where c is a constant.

Hence, as in **12·4**, we have
$$T \cos \psi = T_0$$
and
$$T \sin \psi = \int^s w \, ds,$$
so that
$$\tan \psi = \frac{1}{T_0} \int_0^s w \, ds = \frac{1}{T_0} \int_0^s \frac{T}{c} \, ds = \frac{1}{c} \int_0^s \sec \psi \, ds,$$
and, by differentiation,
$$\sec^2 \psi \frac{d\psi}{ds} = \frac{\sec \psi}{c},$$
or
$$\frac{d\psi}{ds} = \frac{\cos \psi}{c} = \frac{1}{c} \frac{dx}{ds}.$$
Therefore
$$x = c\psi + A,$$
and, if the origin is taken at the lowest point, the constant A is zero, and
$$x = c\psi.$$
Then
$$\frac{dy}{dx} = \tan \psi = \tan \frac{x}{c},$$
so that
$$y = c \log \sec \frac{x}{c}.$$

The curve is symmetrical about the lowest point, and $y \to \infty$ as $x \to \pm \frac{1}{2}\pi c$, so that there are vertical asymptotes and the span cannot exceed πc.

12·6. Elastic Strings. The law relating the tension of an elastic string to its extension was enunciated in **11·8**: but in the cases there considered it was assumed that the tension and extension of all parts of the string were the same. In the case of a heavy string this assumption does not hold, but Hooke's Law is to be applied to each small element of the string.

It is clear that after an elastic string has been stretched and taken up a position of equilibrium, the equations of equilibrium must be of the same form as for an inextensible string, save for the fact that the 'line density' or mass per unit length is the unknown line density which the string possesses after being stretched.

Thus suppose that $m\,\delta s$ is the mass of an element δs of the stretched string, whose line density and length when unstretched are m_0 and δs_0, then

$$m\,\delta s = m_0\,\delta s_0,$$

and by Hooke's Law the tension of the element is

$$T = \lambda \frac{\delta s - \delta s_0}{\delta s_0},$$

therefore

$$m = m_0 \Big/ \left(1 + \frac{T}{\lambda}\right),$$

and this is the additional information required in solving a problem of an elastic string.

12·61. *A heavy elastic string hangs vertically and supports a weight* W.

Let OA represent the stretched string carrying a weight W at A.

Let P be a point at a distance x from O, PQ an element dx, and l the length OA.

Let l_0, x_0, dx_0 be the unstretched lengths of l, x and dx, and let w be the weight of unit length of unstretched string.

Then by Hooke's Law the tension at P is

$$T = \lambda \frac{dx - dx_0}{dx_0} \quad\quad\dots\dots\dots\dots\dots\dots(1).$$

But the tension at P supports the weight W and the weight of the portion PA of the string, so that

$$T = W + w(l_0 - x_0) \quad\quad\dots\dots\dots\dots\dots(2).$$

Therefore, from (1) and (2)

$$dx = dx_0 \left\{1 + \frac{W}{\lambda} + \frac{w}{\lambda}(l_0 - x_0)\right\},$$

and, by integration

$$x = x_0 \left(1 + \frac{W}{\lambda} \right) + \frac{w}{\lambda} (l_0 x_0 - \tfrac{1}{2} x_0{}^2);$$

there being no constant of integration since x and x_0 vanish together.

Putting $x_0 = l_0$, we get for the whole extension

$$\frac{l - l_0}{l_0} = \frac{1}{\lambda} (W + \tfrac{1}{2} w l_0).$$

Also, when W is zero, $\dfrac{l - l_0}{l_0} = \dfrac{w l_0}{2\lambda};$

and the string is doubled in length, i.e. $l = 2l_0$ if $\lambda = \tfrac{1}{2} w l_0 = $ half the weight of the string.

12·62. The Elastic Catenary. An elastic string which is uniform when unstretched hangs between two fixed points.

Using the figure of **12·4** to represent the equilibrium position of the string, and taking w as the weight of unit length of unstretched string, let $CP = s$ and let s_0 be the unstretched length of CP so that $w s_0$ is its weight.

Then by resolving horizontally and vertically for CP we have

$$T \cos \psi = T_0 = wc \quad \text{say} \quad \dots\dots\dots\dots\dots\dots(1)$$

and

$$T \sin \psi = w s_0 \dots\dots\dots\dots\dots\dots\dots(2),$$

so that by squaring and adding

$$T^2 = w^2 (c^2 + s_0{}^2) \dots\dots\dots\dots\dots\dots(3).$$

Also, by Hooke's Law,

$$ds = ds_0 \left(1 + \frac{T}{\lambda} \right) \dots\dots\dots\dots\dots\dots(4).$$

Now $\dfrac{dx}{ds} = \cos \psi = \dfrac{wc}{T}$, from (1);

therefore $x = wc \displaystyle\int \frac{ds}{T} = wc \int \left(\frac{1}{T} + \frac{1}{\lambda} \right) ds_0$, from (4)

$$= \int \left\{ \frac{c}{\sqrt{(c^2 + s_0{}^2)}} + \frac{wc}{\lambda} \right\} ds_0, \text{ from (3)}$$

$$= c \sinh^{-1} \frac{s_0}{c} + \frac{wc s_0}{\lambda} \dots\dots\dots\dots\dots\dots(5),$$

with no constant of integration if the axis of y passes through the lowest point of the string.

Similarly $\dfrac{dy}{ds} = \sin \psi = \dfrac{w s_0}{T}$, from (2).

Therefore $y = w \displaystyle\int \frac{s_0}{T} ds = w \int \left\{ \frac{s_0}{\sqrt{(c^2 + s_0{}^2)}} + \frac{w s_0}{\lambda} \right\} ds_0$, from (3)

$$= \sqrt{(c^2 + s_0{}^2)} + \frac{w s_0{}^2}{2\lambda} \dots\dots\dots\dots\dots\dots\dots(6),$$

with no constant of integration if the axis of x is at a depth c below the lowest point of the string.

Equations (5) and (6) express the co-ordinates of a point on the curve in terms of the parameter s_0.

12·7. Miscellaneous Examples. (i) *A rope is passed round a framework of rectangular beams with rounded corners as shewn in the figure. If the diameter of the rope d is small compared with the dimensions of the*

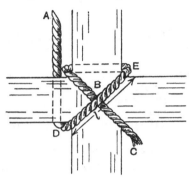

framework, shew that the condition that it should not slip when a tension is applied at A, the end C being free, is $\mu e^{2\pi\mu} > l/8d$, where l is measured as in the figure and μ is the coefficient of friction both between the rope and the beams and between the two parts of the rope.

[*Assume that the frictions at B act along BC.*]   [T.]

Let R be the pressure at B between the framework and the under rope and also between the two ropes.

Let T and T' be the tensions in the under and upper ropes at B.

The upward pressure R on an element of the upper rope at B must be balanced by the resolved

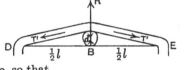

part of the tensions T' on either side, so that

$$R = 2T'\frac{d}{l/2} = 4\frac{d}{l}T'.$$

The friction available to prevent the sliding of the lower rope in the direction CB is then $2\mu R$ or $\dfrac{8\mu d}{l}T'$, and this must exceed the tension T in order to maintain equilibrium.

But in passing along the rope, from the lower to the upper position at B, there are two turns in the rope at each of which the rope turns through an angle $\frac{1}{2}\pi$ in getting at right angles to the plane of the figure and through another $\frac{1}{2}\pi$ in getting back into the plane of the figure, so that the total angle turned through is 2π, and, by **12·12**, $T' = Te^{2\mu\pi}$.

Therefore the condition that the rope should not slip is

$$\frac{8\mu d}{l}e^{2\mu\pi} > 1,$$

or
$$\mu e^{2\mu\pi} > l/8d.$$

(ii) *Prove that in order that a chain of variable density may hang in a given curve, the weight per unit of the horizontal co-ordinate x must be* $T_0\,d^2y/dx^2$, *where* T_0 *is the horizontal tension.*

A chain whose ends are fixed at A and B revolves about AB with constant angular velocity (gravity neglected). Prove that if it has the form of a curve of sines having points of zero curvature at A and B, the mass of any portion must be proportional to the orthogonal projection of that portion on the line AB. [T.]

For the first part, using the figure of **12·4** and taking w as the weight per unit length of the horizontal co-ordinate x, the weight of the portion CP is $\int_0^x w\,dx$, so that

$$T\cos\psi = T_0,$$

and
$$T\sin\psi = \int_0^x w\,dx.$$

Therefore
$$T_0\tan\psi = \int_0^x w\,dx,$$

or
$$T_0\frac{dy}{dx} = \int_0^x w\,dx;$$

whence, by differentiation $w = T_0\dfrac{d^2y}{dx^2}.$

For the second part let $y = a\sin x$ be the curve of the chain, with A as origin and AB as axis of x.

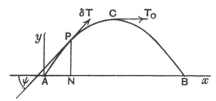

Let P be the point (x, y) and $m\,\delta s$ the mass of a small element at P. This has an acceleration $\omega^2 y$ along the ordinate PN, and the forces producing this acceleration are the tangential and normal components of the tension at the ends of the element, i.e. δT along the tangent and $T\,\delta\psi$ along the inward normal. Hence, by resolving along the tangent, we get $m\,\delta s\,.\,\omega^2 y\sin\psi = -\,\delta T.$

Also, by resolving parallel to AB for the portion CP we have

$$T\cos\psi = T_0.$$

Therefore the mass of the portion AP is

$$\int_0^s m\,ds = -\int \frac{dT}{\omega^2 y \sin \psi} = -\frac{T_0}{\omega^2}\int \frac{\sec^2 \psi}{y}\,d\psi,$$

where $\qquad\qquad y = a\sin x$, so that $\tan \psi = \dfrac{dy}{dx} = a\cos x$,

and $\qquad\qquad\qquad \sec^2 \psi\,d\psi = -a\sin x\,dx.$

Therefore $\qquad\qquad \int_0^s m\,ds = \dfrac{T_0}{\omega^2}\int_0^x dx = \dfrac{T_0 x}{\omega^2}.$

(iii) *An extensible string, uniform when unstretched and of length l, lies initially unstretched in a straight line on a rough horizontal plane. The string is then pulled at one end in the direction of its line produced, with a gradually increasing force, so that its acceleration is always infinitely small. Prove that when the force is F, the extension of the string is $\frac{1}{2}F^2 l/\mu W\lambda$, where W is the weight of the string, λ the coefficient of elasticity, μ the coefficient of friction, and $F < \mu W$.* [T.]

$$\overset{\text{L}}{\vert}\text{———————}\overset{\text{N}}{\vert}\text{————}\overset{\text{P}}{\vert}\text{———}\overset{\text{O}}{\vert}\text{———}\overset{\text{F}}{\longrightarrow}$$

Let OL represent the string when the force is F, and let x denote the distance of any point P from the end O at which F is applied. Let x_0 be the unstretched length of the portion OP and T the tension at P. The weight of OP is $\dfrac{x_0}{l}W$ and the friction acting on it is $\dfrac{\mu x_0}{l}W$, so that

$$T + \frac{\mu x_0}{l}W = F \qquad\qquad\dots\dots\dots\dots\dots\dots(1).$$

Now since $F < \mu W$, there is not enough force to move the whole string, so we assume that when the force is F a certain portion ON has been extended while NL remains unstretched and this implies that there is a point N at which the tension vanishes. The unstretched length of ON is obtained from (1) by putting $T = 0$, which gives

$$x_0 = lF/\mu W \qquad\qquad\dots\dots\dots\dots\dots\dots(2).$$

Again, by Hooke's Law, the tension at P is given by

$$T = \lambda \frac{dx - dx_0}{dx_0},$$

so that from (1) $\qquad dx - dx_0 + \dfrac{\mu W}{\lambda}x_0\,dx_0 = \dfrac{F}{\lambda}dx_0,$

and, by integration,

$$x - x_0 + \frac{1}{2}\frac{\mu W}{\lambda}x_0^2 = \frac{F}{\lambda}x_0 \qquad\dots\dots\dots\dots\dots(3),$$

there being no constant of integration since x and x_0 vanish together at O.

The whole extension when the force is F is then the value of $x - x_0$ when for x_0 we take the unstretched length of ON given by (2), and we have

$$x - x_0 = \frac{F}{\lambda} x_0 - \frac{1}{2} \frac{\mu W}{\lambda} x_0{}^2$$

$$= \frac{F}{\lambda} \cdot \frac{lF}{\mu W} - \frac{1}{2} \frac{\mu W}{\lambda} \cdot \frac{l^2 F^2}{\mu^2 W^2}$$

$$= \tfrac{1}{2} F^2 l / \mu W \lambda.$$

EXAMPLES

1. A rope is coiled round two fixed bollards as shewn in the figure, and one end is held with a force of 50 lb. Find the greatest force which can be applied at the other end without causing the rope to slip. Take the coefficient of friction between the rope and the bollards to be 0·2.
[S.]

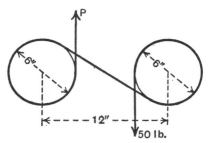

2. Three equally rough pegs A, B, C of the same circular cross section are placed at the corners of an equilateral triangle of side a, so that BC is horizontal and A above BC. Shew that the greatest weight which can be supported by a weight W tied to the end of a string, which is carried once round the pegs and never completely surrounds any peg, is $We^{5\mu\pi}$, μ being the coefficient of friction. [C.]

3. A uniform chain rests on the upper half of a smooth vertical circle, its ends reaching to the horizontal diameter. Prove that the pressure per unit length at the highest point is twice the weight per unit length.

4. A uniform chain hangs round a smooth vertical circle. Prove that if the pressure vanishes at the lowest point, then the tension at the highest point is three times that at the lowest point.

5. A uniform chain hangs beneath the lower half of a smooth vertical circle. Prove that if the pressure per unit length at the lowest point is equal to the weight per unit length the tension at the ends must be half as much again as the tension at the lowest point.

6. A uniform string rests in limiting equilibrium on a rough quadrant of a circle with one end at the lowest point. Prove that the angle of friction λ satisfies the relation

$$\tfrac{1}{2}\pi \tan \lambda = \log \tan 2\lambda.$$

7. A uniform string rests on a smooth cycloidal curve whose axis is vertical and vertex upwards, the string just reaching to the cusps of the cycloid. Prove that the pressure at any point varies as the curvature.

8. If the tangents at the points P and Q of a catenary are at right angles, prove that the tension at the middle point of the arc PQ is equal to the weight of a length of the string equal to half the arc PQ.

9. A uniform string of weight W rests on a rough cycloidal curve which has its axis vertical and vertex upwards, the string extending from the vertex to a cusp. Shew that the least horizontal force applied at the vertex that will cause the string to slip upwards is

$$W (3e^{\mu\pi/2} - \mu^2 - 1)/(\mu^2 + 4),$$

where μ is the coefficient of friction.

10. A uniform chain is hung up by its two ends which are on the same level and the sag in the middle is small; shew that the terminal tension varies inversely as the sag. [S.]

11. A chain 40 ft. long, which weighs 1 lb. per foot, hangs between two points on the same level and has a sag of 5 ft. Find the parameter of the catenary and the tension at a point of support. [I.]

12. A telegraph wire, stretched between two poles at distance a feet apart, sags n feet in the middle; prove that the tension at the ends is approximately

$$w\left(\frac{a^2}{8n} + \frac{7}{6}n\right).$$ [I.]

13. ACB is a telegraph wire, the straight line AB being horizontal and of length $2l$, and C the middle point of the wire is at a distance h below AB. Shew that the length of the wire is approximately

$$2l + \frac{4}{3}\frac{h^2}{l} - \frac{28}{45}\frac{h^4}{l^3}.$$ [S.]

14. A chain of length 20 ft. and weight 10 lb. is stretched nearly straight between two points at different levels. Assuming that vertically below the middle point of the chord the chain is approximately parallel to the chord and that the tension there is 100 lb. weight, prove that the sag measured vertically from the middle point of the chord is approximately 3 in. [S.]

15. A suspension chain carries a load uniformly distributed on a horizontal platform. The load is $\frac{1}{2}$ ton per foot length of the span of 600 ft., and the height of the point of support above the lowest point of the chain is 50 ft. Find the greatest and least tensions in the chain, neglecting its weight. [T.]

16. The span of a suspension bridge is 100 ft. and the sag at the middle of the chain is 10 ft.: if the total load on each chain is 25 tons, find the greatest tension in each chain and the tension at the lowest point. [S.]

17. A telegraph wire, of length l, hangs between two posts on the same level, at distance a apart, the small sag in the centre being b; shew that $l - a = \frac{8}{3} b^2/a$, approximately, and that the least tension per unit area of the section is $\frac{1}{4} wa^2/b$, if w is the weight per unit volume of the wire.

If the greatest tension is to be 15,000 lb. per square inch of section, calculate the least sag allowable in a span of 100 yd. if the wire weighs ·28 lb. per cubic inch; and prove that the increase in length due to the sagging is about 1 ft. per mile. [C.]

18. If a telegraph wire has a sag of 30 in. in a span of 100 yd., calculate the least tensile stress per square inch of the section, assuming the weight to be ·28 lb. per cubic inch.

What extra length of wire would be required over 10 miles, supposing the posts uniformly spaced at 100 yd. apart and the sag to be 30 in. in each span? [C.]

19. In a suspension bridge of 600 ft. span and 60 ft. dip the whole weight supported is 2 tons per horizontal foot run. Find the horizontal tension in each of the two chains and the tension in each at the points of support.

A freely hanging chain of the same span carries the same load per foot of its own length and has the same horizontal tension as in the case above; find its dip and the tension at the points of support.

$$[\cosh \cdot 8 = 1 \cdot 337.] \qquad [\text{T.}]$$

20. A box-kite is flying at a height h, with a length l of wire paid out, and with the vertex of the catenary on the ground; prove that at the kite $\tan \frac{1}{2} \psi = h/l$, and that the tension there is equal to the weight of a length $\frac{1}{2} (l^2 + h^2)/h$ of wire. [C.]

21. A heavy uniform string rests on a smooth catenary with its axis vertical and its vertex upwards; prove that the pressure on the curve at any point varies inversely as the square of the distance of that point below the directrix. [I.]

22. Two smooth rods are situated in the same vertical plane and make equal angles α with the downward vertical. A uniform chain of length l hangs by weightless rings from the rods. Prove that the distance between the rings is

$$l \cot \alpha \sinh^{-1} \tan \alpha. \qquad [\text{I.}]$$

23. A particle of mass m is suspended from a fixed point by a light string which is blown from the vertical by a steady horizontal wind of uniform velocity V. Assuming that the force exerted by the wind on each element of length δs of the string is normal to the element and of

magnitude $\kappa v^2 \delta s$, v being the component of V normal to the element, shew that the string hangs in a catenary.

If the wind pressure on the particle is negligible, prove that the depth of the particle below the point of suspension is

$$c \log \left[\frac{l + \sqrt{l^2 + c^2}}{c} \right],$$

where $c = mg/\kappa V^2$, and l is the length of the string. [S.]

24. Prove that if a uniform chain lie in one plane under the action of a system of force fixed in direction and of magnitude at any point varying as $\operatorname{cosec}^2 \psi$, where ψ is the inclination of the chain to the fixed direction, the chain must lie along an arc of a circle. [I.]

25. A heavy string of length $2l$ is hung from two fixed points A, B in the same horizontal line, at a distance apart equal to $2a$. A weight W is attached to a certain point of the string. Shew that the parameters of the two catenaries in which the string hangs are the same, and shew that if W is in the middle of the string, and its weight is great in comparison to that of the string, the parameter c is equal to $Wa/2w\sqrt{l^2 - a^2}$ nearly, while if on the other hand the weight of W is small in comparison with that of the string,

$$l = c \sinh \frac{a}{c} + \frac{W}{2w} \left\{ \cosh \frac{a}{c} - 1 \right\},$$

w being the weight per unit length of the string. [S.]

26. A chain hangs freely in the form of an arc of a circle. Shew that its weight per unit length at any point varies as the square of the secant of the angle which the radius to that point makes with the vertical. [S.]

27. If a heavy uniform string hangs in equilibrium over two smooth pegs, prove that the free ends must be at the same level.

If the pegs are rough and in the same horizontal line and every point of the string is on the point of slipping in the direction of the corresponding tangent, prove that the lengths of the vertical portions are in the ratio $e^{\mu(\pi+2\psi)} : 1$, where μ is the coefficient of friction and $\psi = \tan^{-1} \dfrac{l}{c}$, $2l$ being the length hanging between the pegs and c the parameter of the catenary. [I.]

28. A heavy string of variable density is hung up from two points. Prove that if T_1, T_2, T_3 are the tensions at points A, B, C, where the inclinations of the tangents to the horizon are $\alpha - \beta$, α, $\alpha + \beta$, and w_1, w_2 the weights of the parts of the string AB, BC respectively,

$$\frac{1}{T_1} + \frac{1}{T_3} = \frac{2 \cos \beta}{T_2},$$

$$\frac{T_1}{w_1} = \frac{T_3}{w_2}.$$

[S.]

29. In any network of strings hanging in a vertical plane, if three uniform inextensible strings of the same line density meet in a knot and their directions at that point make angles θ, ϕ, ψ with the horizontal, prove that the parameters of the catenaries are numerically as

$$\tan\phi - \tan\psi : \tan\psi - \tan\theta : \tan\theta - \tan\phi. \qquad \text{[I.]}$$

30. A ring of weight wb is attached to the middle point of a string of length l which hangs symmetrically over two smooth pegs in the same horizontal line, the ends of the string being vertical. If w is the weight per unit length of the string and $2a$ the distance apart of the pegs, prove the equation to determine the parameter of the catenaries in which the string hangs, $b + l = e^{a/c}\{b + \sqrt{4c^2 + b^2}\}$; and shew that the least value of l for which equilibrium is possible occurs when

$$1/(c-a)^2 - 1/c^2 = 4/b^2. \qquad \text{[I.]}$$

31. One end of a uniform rough string is fastened to a point P, at a height h above a table, and part of the string rests on the table in a vertical plane through P. Shew that the greatest length which can lie on the table is the smaller root of the equation

$$z^2 - 2(l + \mu h)z + l^2 - h^2 = 0,$$

where l is the length of the string. [I.]

32. One end of a rough uniform chain of length l is fastened to a point on a vertical wall at a height h above the ground. Shew that the greatest distance from the wall at which the free end of the chain will rest on level ground is given by the expression

$$u\left\{1 + \mu\log\left(\frac{h+l}{\mu u} + 1 - \frac{1}{\mu}\right)\right\},$$

where $\qquad u = l + \mu h - \{(\mu^2 + 1)h^2 + 2\mu lh\}^{\frac{1}{2}}$,

and μ is the coefficient of friction. [S.]

33. A weightless string is attached to two points in the same vertical line at a distance h apart. The wind produces a force $k\sin\psi$ per unit length in the direction of the normal, where ψ is the angle between the tangent to the string and the horizontal, and k is a constant. Prove that the tension T is the same at all points of the string and that it cannot be less than kh/π however long the string may be. [T.]

34. A uniform chain of length l and weight W hangs between two fixed points at the same level and a weight is suspended from its middle point so that the total sag in the middle is h. Shew that, if P is the pull on either point of support, the total load is

$$\frac{4h}{l}P + \left(\frac{1}{2} - \frac{2h^2}{l^2}\right)W.$$

35. Prove that in the catenary of uniform strength

$$\rho = c\sec\psi = c\cosh\frac{s}{c}.$$

36. One end A of a uniform string of length l is fixed, the other end B moves along a horizontal line through A. Shew that the locus of the vertex of the catenary formed by the string referred to horizontal and vertical axes through A is given by

$$8xy = (l^2 - 4y^2) \log \frac{l + 2y}{l - 2y}. \qquad [\text{C.}]$$

37. A uniform telegraph wire is made of a given material and such a length l is stretched between two posts distant d apart as will produce the least possible stress on the posts. Shew that $l = d/\lambda \sinh \lambda$, where λ satisfies the equation $\lambda \tanh \lambda = 1$. It may be assumed that the ends of the wire are at the same level. [C.]

38. In a non-uniform string hanging under gravity the area of the cross section at any point is inversely proportional to the tension. Prove that the curve is an arc of a parabola with its axis vertical. [I.]

39. Find the intrinsic equation of the catenary formed by an elastic thread which is uniform when unstretched. Shew that for a large modulus it approximates to a common catenary, and for a small modulus to a parabola. [I.]

ANSWERS

1. 405 lb.   11. 37·5 ft.; 42·5 lb.
15. 474·3 tons, 450 tons.   16. 33·66 tons, 31·25 tons.
17. 30 in.   18. 15,120 lb.; 9·9 ft.
19. 750 tons, 807·75 tons, 126·5 ft., 1003 tons.

39. $s = c \tan \psi + \dfrac{wc^2}{2\lambda} \{\tan \psi + \sec \psi + \log (\tan \psi + \sec \psi)\}.$

Chapter XIII

ELASTICITY

13·1. So far we have been dealing largely with such mathematical fictions as rigid bodies, weightless rods or inextensible strings, but in the present chapter we propose to bring our investigations into closer touch with reality by showing how to make allowance for the fact that bodies are not rigid but undergo small changes in form when subject to the action of force.

We shall confine our considerations to a few simple cases of *isotropic* bodies. An isotropic body is such that if a sphere is cut out of the body anywhere it possesses no directional properties of any kind, as distinct from a crystalline body or a body of fibrous structure.

The simplest type of deformation or strain that a body can undergo is a uniform extension, in which *all* elements of length PQ in a certain direction are altered to elements of length $P'Q'$ such that the increment in length $P'Q' - PQ$ is a certain fraction ϵ of the original length, i.e. $\epsilon = (P'Q' - PQ)/PQ$. This fraction ϵ is then called the *extension*.

Regarding a contraction as a negative extension it is clearly possible for a body to be extended in more directions than one. It is obvious, for example, that if a bar is extended longitudinally it will in general contract laterally. We do not propose, however, to analyse the different kinds of strain that are possible but only to deal with some simple cases.

13·2. Extension of Bars. Young's Modulus. Within certain limits the extension or compression of a uniform bar is related to the tension or thrust in it by Hooke's Law (**11·8**), i.e.

$$T = \lambda \frac{l - l_0}{l_0},$$

where λ is called the modulus of elasticity of the bar. It is, however, more usual to put $\lambda = E\omega$, where ω is the area of the cross section of the bar and E is a constant known as Young's modulus of elasticity for the material of which the bar is composed. If we write ϵ for the extension $(l - l_0)/l_0$, we have

$$T/\omega = E\epsilon,$$

or the stress per unit area of the cross section is E times the extension.

We assume as above that Hooke's Law is true, as a first approximation at any rate, for small extensions or compressions of bar or other bodies, i.e. that if equal and opposite forces are applied longitudinally at the ends of a bar, extension or compression will take place in accordance with this law and that when the forces are removed the body returns to its former state. But there is a limit in each case to the magnitude of the displacement for which the law remains true, known as the *elastic limit* or the *limit of linear elasticity*, and we must assume that these limits are not exceeded. There are materials, such as cast-iron and cement, which do not obey Hooke's Law at all.

13·21. Variable Extension. When there are externally applied forces the stress per unit area along a bar generally varies from one cross section to another so that the extension is also variable.

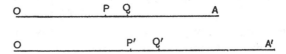

Let O be a fixed point on the bar OA when unstressed. Let $OP = x$ and $PQ = \delta x$. Let O, P', Q' A' be the points which correspond to O, P, Q, A in the strained condition, and let $OP' = x + \xi$, so that ξ denotes the displacement of the point P.

Then
$$OQ' = x + \delta x + \xi + \frac{\partial \xi}{\partial x}\delta x,$$

so that
$$P'Q' = \left(1 + \frac{\partial \xi}{\partial x}\right)\delta x.$$

Therefore the extension is given by

$$\epsilon = \frac{P'Q' - PQ}{PQ} = \frac{\partial \xi}{\partial x},$$

and if ω is the cross section, the total stress in the bar at P is

$$T = E\omega\epsilon = E\omega \frac{\partial \xi}{\partial x}.$$

13·22. Extension of a Bar under its own weight. This is the same problem as that of the extension of an elastic string in **12·61.**

With our present notation and w for the weight of unit volume of the bar, l for the total length and T for the tension at a point whose unstrained distance from the upper end is x, we have

$$T = w\omega(l - x).$$

Therefore

$$E \frac{\partial \xi}{\partial x} = w(l - x),$$

and

$$E\xi = w(lx - \tfrac{1}{2}x^2);$$

no constant of integration being required since ξ and x vanish together.

The total increment in length of the bar is the value of ξ when $x = l$, i.e.

$$\tfrac{1}{2}wl^2/E.$$

13·23. Elastic Energy of a Longitudinal Strain. Consider a unit cube of a body which is to be subject to a longitudinal strain in the direction of one set of edges. Suppose that at any

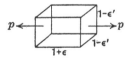

instant these edges have increased in length to $1 + \epsilon$ under a longitudinal stress of magnitude p per unit of area. Assuming there to be no change of volume the other edges of the cube will have contracted to $1 - \epsilon'$, such that $(1 + \epsilon)(1 - \epsilon')^2 = 1$, or to the first order of ϵ and ϵ', $\epsilon - 2\epsilon' = 0$, so that $\epsilon' = \tfrac{1}{2}\epsilon$.

The stress p per unit area is connected with the extension by the formula $p = E\epsilon$, and the work done in a further small extension of the cube is

$$p(1 - \tfrac{1}{2}\epsilon)^2 d\epsilon, \quad \text{or} \quad E\epsilon(1 - \tfrac{1}{2}\epsilon)^2 d\epsilon,$$

so that the work done in the extension from 0 to ϵ is

$$\int_0^\epsilon E\epsilon(1-\tfrac{1}{2}\epsilon)^2\,d\epsilon$$

or $\tfrac{1}{2}E\epsilon^2$, neglecting higher powers of ϵ.

This is a measure of the elastic energy per unit volume assuming that there is free (i.e. unopposed) lateral contraction.

13·3. Bending of Bars. Consider a straight bar to be slightly bent in such a way that plane cross sections remain plane sections but are no longer parallel but slightly inclined to one another.

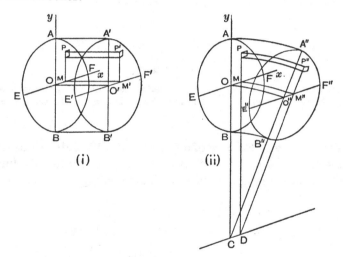

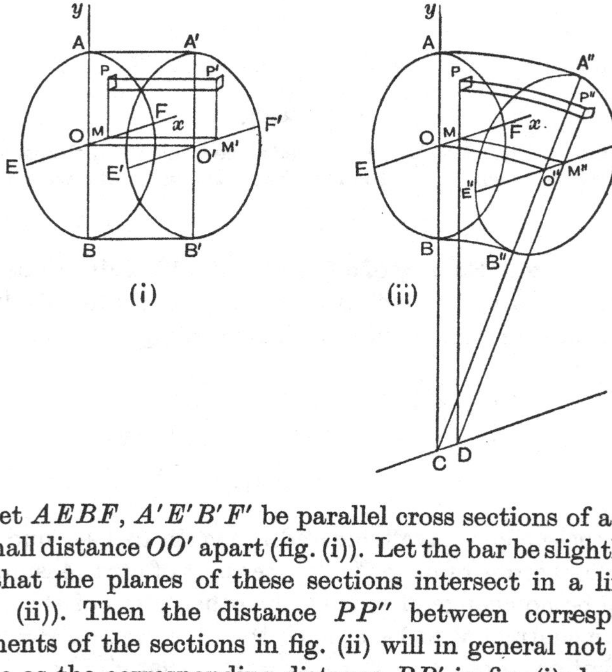

(i)   (ii)

Let $AEBF$, $A'E'B'F'$ be parallel cross sections of a bar at a small distance OO' apart (fig. (i)). Let the bar be slightly bent so that the planes of these sections intersect in a line CD (fig. (ii)). Then the distance PP'' between corresponding elements of the sections in fig. (ii) will in general not be the same as the corresponding distance PP' in fig. (i): but there will be a pair of lines EF, $E''F''$ parallel to CD in fig. (ii) whose distance apart has not been altered by the bending. These are called the neutral lines. Take rectangular axes Ox, Oy in the plane $AEBF$ with Ox along EF; and consider a narrow strip

PP' of the bar of cross section $dxdy$, which becomes PP'' when the bar is bent; P being the point (x, y). The extension ϵ of this strip is given by

$$\epsilon = \frac{PP'' - PP'}{PP'}.$$

But from fig. (i) $PP' = MM'$, and MM' is unaltered by bending and is equal to MM''; also

$$\frac{PP''}{MM''} = \frac{PD}{MD} \quad \text{(fig. (ii))}$$

$$= \frac{y + R}{R},$$

where for MD we write R, the radius of curvature of the curve MM'' or OO''.

Therefore $\epsilon = \dfrac{y}{R}$, and the stress in this strip of cross section $dxdy$ is

$$\frac{Ey}{R}\,dxdy.$$

Hence the resultant stress in the bar taken over the cross section $AEBF$ is

$$\frac{E}{R} \iint y\,dxdy,$$

and this will vanish if the centroid of the cross section lies in the neutral line Ox.

It follows that if a bar is unstressed (i.e. if the resultant stress across every cross section is zero) but slightly bent by the application of couples, the centroids of the cross sections lie on a central line OO', which undergoes no extension in the bending.

Consider next what are the sums of the moments about the axes Oy and Ox of the stresses over the cross section $AEBF$.

The resultant moment about Oy is

$$\frac{E}{R} \iint xy\,dxdy,$$

and this will vanish if the axis Oy is a principal axis of inertia of the cross section or simply an axis of symmetry of the cross section. In this case the couples necessary to maintain equilibrium will lie entirely in the plane of the bending; the total moment of the stresses about Ox being

$$\frac{E}{R} \iint y^2 \, dx \, dy$$

or $\frac{EI}{R}$, where I is the moment of inertia of the area of the cross section about the neutral line, and R is the radius of curvature of the central line of the bar.

This expression EI/R measures the *bending moment* at any cross section of the bar, as defined in Chapter VII.

13·31. In using the results of **13·3** it must be remembered that they have only been obtained with certain assumptions, e.g. that plane sections remain plane sections, and without taking account of the local effects of concentrated loads or of points of support, and that when we make use of these results to deduce others we must accept them with reservations as to how far they will necessarily accord with the results of experiment.

13·4. Applications. In Chapter VII we investigated the method of measuring the shearing force and bending moment at any point of a beam in an equilibrium position. We have in **13·3** obtained a measure of the bending moment in a strained position of a beam in terms of its shape in this position and Young's modulus for the material of which it is composed. We shall now show how, by equating these two measures of the bending moment, to determine the strained form of the beam on the hypothesis that the strain is small.

We shall assume that the breadth and depth of the beam are small compared to its length, and that the deviation from the unstrained state is small, so that if we take an axis of x along the unstrained direction of the beam it will be sufficient to take d^2y/dx^2 as the measure of the curvature $1/R$ of the bent beam. We may also use a single symbol B to denote the product EI when the beam is of uniform cross section; the bending moment is then $B \dfrac{d^2y}{dx^2}$ and B is called the *flexural rigidity* of the beam.

13·41. Care must be taken about the sign of the bending moment. Thus, if, in the figure, AB is a beam divided at P, with the axes as shewn, then $B\dfrac{d^2y}{dx^2}$ at P

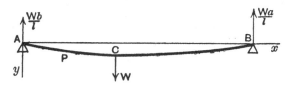

= clockwise moment about P of the forces acting upon PB, or

= counter-clockwise moment about P of the forces acting upon PA.

13·42. Examples. (i) *A light beam of length l is supported at its ends at the same level and carries a load W concentrated at a point at a distance a from one end.* [7·21 (i).]

Let AB be the beam,

$$AC=a,\quad CB=b=l-a.$$

The forces of support at A and B are Wb/l and Wa/l. Taking an axis of x along AB and the axis of y vertically downwards at A, let P be the point (x, y), then taking the bending moment at P, we have, for

$$0<x<a,\quad B\frac{d^2y}{dx^2}=-\frac{Wb}{l}x,$$

so that

$$B\left(\frac{dy}{dx}-\tan\alpha\right)=-\frac{1}{2}\frac{Wb}{l}x^2 \quad\ldots\ldots\ldots\ldots(1),$$

where $\tan\alpha$ is the gradient of the central line at A; and

$$B(y-x\tan\alpha)=-\frac{1}{6}\frac{Wb}{l}x^3 \quad\ldots\ldots\ldots\ldots(2),$$

with no further constant of integration, since x and y vanish together.

Again, for $a<x<l,\quad B\dfrac{d^2y}{dx^2}=-\dfrac{Wa}{l}(l-x),$

so that

$$B\left(\frac{dy}{dx}+\tan\beta\right)=\frac{1}{2}\frac{Wa}{l}(l-x)^2\ldots\ldots\ldots\ldots\ldots(3),$$

where β is the acute angle which the beam makes with the horizontal at B; and

$$B\{y-(l-x)\tan\beta\}=-\frac{1}{6}\frac{Wa}{l}(l-x)^3\ldots\ldots\ldots\ldots(4),$$

with no further constant of integration since y vanishes when $x=l$.

Now there is no discontinuity at C where $x=a$, so that, when $x=a$, (1) and (3) will give the same value for dy/dx and (2) and (4) will give the same value for y.

Therefore $\quad B\tan\alpha - \dfrac{1}{2}\dfrac{Wba^2}{l} = -B\tan\beta + \dfrac{1}{2}\dfrac{Wab^2}{l},$

and $\quad Ba\tan\alpha - \dfrac{1}{6}\dfrac{Wba^3}{l} = Bb\tan\beta - \dfrac{1}{6}\dfrac{Wab^3}{l};$

whence $\qquad B\tan\alpha = \dfrac{1}{6}\dfrac{W}{l}ab(a+2b),$

and $\qquad B\tan\beta = \dfrac{1}{6}\dfrac{W}{l}ab(2a+b).$

Then the deflection at different points of the beam is given from (2) and (4) thus

$$0<x<a, \qquad By = \dfrac{1}{6}\dfrac{Wb}{l}\{a(a+2b)x - x^3\} \quad\dots\dots\dots\dots(5),$$

and $a<x<l, \qquad By = \dfrac{1}{6}\dfrac{Wa}{l}\{b(2a+b)(l-x)-(l-x)^3\} \quad\dots(6).$

As a special case when the load is at the middle point of the beam, so that $a=b=\frac{1}{2}l$, we have,

$$0<x<\tfrac{1}{2}l, \qquad By = \tfrac{1}{12}W(\tfrac{3}{4}l^2 x - x^3),$$

and $\tfrac{1}{2}l<x<l, \qquad By = \tfrac{1}{12}W\{\tfrac{3}{4}l^2(l-x)-(l-x)^3\};$

and the deflection at the middle point is $\tfrac{1}{48}Wl^3/B$.

(ii) *A light beam is clamped horizontally at one end and carries a load W concentrated at the other end.*

Let l be the length of the beam AB clamped at A.

Taking horizontal and vertical axes through A, we have for the bending moment at a point $P(x,y)$

$$B\dfrac{d^2y}{dx^2} = W(l-x),$$

so that $\quad B\dfrac{dy}{dx} = W(lx - \tfrac{1}{2}x^2),$

no constant being necessary because, by hypothesis, dy/dx vanishes when $x=0$.

Integrating again, we get

$$By = W(\tfrac{1}{2}lx^2 - \tfrac{1}{6}x^3),$$

since x and y vanish together.

We find for the deflection at B, where $x=l$,

$$y = \tfrac{1}{3}Wl^3/B.$$

The result could have been predicted from the preceding example. For if we fixed the ends of a beam of length $2l$ at the same level and applied an upward force $2W$ at the middle point, the beam would be horizontal in the middle and the deflection at the middle would, by

Ex. (i), be $\frac{1}{4}Wl^3/B$. But the downward force at each end would be W, and either half of the beam might be regarded as clamped at one end and carrying a load W at the other.

We solved this problem by taking moments about P for the forces acting upon PB. We might obtain the same result by taking moments about P for the forces acting upon AP; but in this case we must note that the clamping of the beam at A means that there is a bending moment at A as well as a supporting force W. If we represent this as a couple M (counter-clockwise as it acts upon AB), we get by moments about A for the whole beam

$$M - Wl = 0.$$

Then, by moments about P for the forces on AP, we have

$$B\frac{d^2y}{dx^2} = M - Wx,$$

or

$$B\frac{d^2y}{dx^2} = W(l-x),$$

as before.

13·43. In the problems considered in **13·42** and in kindred problems the equations for the deflections are linear and therefore solutions can be superposed and results obtained by addition. For example, if we consider the case of a uniform heavy beam supported at its ends we may add the deflection to that obtained in Ex. (i) caused by a concentrated load on a beam of negligible weight.

Uniform heavy beam of length l with supports at the same level at its ends.

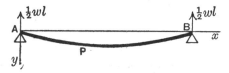

If w denotes the weight of unit length, the supporting forces are each $\frac{1}{2}wl$, and for the bending moment at a point P (x, y) we have

$$B\frac{d^2y}{dx^2} = \frac{1}{2}wx^2 - \frac{1}{2}wlx,$$

so that

$$B\left(\frac{dy}{dx} - \tan\alpha\right) = \frac{1}{6}wx^3 - \frac{1}{4}wlx^2 \quad\text{...............}(1),$$

where $\tan\alpha$ is the gradient at $x = 0$.

Also

$$B(y - x\tan\alpha) = \frac{1}{24}wx^4 - \frac{1}{12}wlx^3 \quad\text{..............}(2),$$

since x and y vanish together.

But y vanishes when $x = l$, therefore

$$B \tan \alpha = \tfrac{1}{24} w l^3 \quad \dots\dots\dots\dots\dots\dots(3),$$

and therefore, from (2),

$$By = \tfrac{1}{24} w x (l - x)(l^2 + lx - x^2) \quad \dots\dots\dots\dots(4).$$

This makes the sag at the middle point $\tfrac{5}{384} w l^4 / B$.

13·44. It is sometimes advantageous to make use of the relations between shearing force and bending moment established in **7·3**; e.g. for a beam carrying a uniformly distributed load w per unit length, the shearing force S and bending moment M satisfy the relations

$$\frac{dS}{dx} = -w \quad \text{and} \quad \frac{dM}{dx} = -S.$$

But $\qquad M = B \dfrac{d^2 y}{dx^2}$, so that $B \dfrac{d^3 y}{dx^3} = -S$,

and $B \dfrac{d^4 y}{dx^4} = w$, but subject to the conditions that at a free end $M = 0$ and $S = 0$, i.e. $\dfrac{d^2 y}{dx^2} = 0$ and $\dfrac{d^3 y}{dx^3} = 0$; whereas at a clamped end there is a bending moment, but the direction being fixed we have $\dfrac{dy}{dx} = 0$ instead of $\dfrac{d^2 y}{dx^2} = 0$.

13·45. The Theorem of Three Moments. Let a uniformly loaded beam be supported at a number of points at the same level and let M_A, M_B, M_C denote the bending moments at

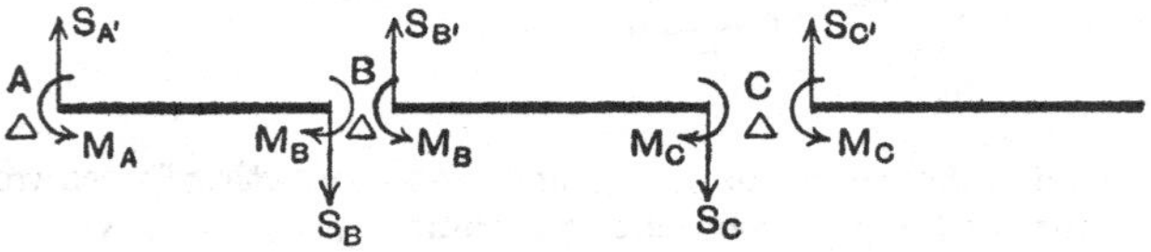

three consecutive points of support A, B, C. Let the shearing forces be as indicated in the diagram, taking note that there will be a discontinuity in the shearing force at each point of support, and that the pressure on the support B, say, will be

$$S_{B'} - S_B.$$

Let $AB = l$, $BC = l'$, and let w be the load per unit length.

By taking moments about B and C for AB and BC respectively, we get

$$lS_{A'} + M_B - M_A - \tfrac{1}{2}wl^2 = 0, \quad \text{and} \quad l'S_{B'} + M_C - M_B - \tfrac{1}{2}wl'^2 = 0$$
$$\dots\dots(1).$$

Then for the bending moment at a point in AB at a distance x from A we have

$$B\frac{d^2y}{dx^2} = \tfrac{1}{2}wx^2 + M_A - S_{A'}x,$$

so that

$$B\left(\frac{dy}{dx} - \tan\alpha\right) = \tfrac{1}{6}wx^3 + M_A x - \tfrac{1}{2}S_{A'}x^2 \quad\dots\dots(2),$$

where $\tan\alpha$ is the value of dy/dx at A; and

$$B\,(y - x\tan\alpha) = \tfrac{1}{24}wx^4 + \tfrac{1}{2}M_A x^2 - \tfrac{1}{6}S_{A'}x^3 \quad\dots\dots(3),$$

since x and y vanish together at A.

But y also vanishes when $x = l$, therefore

$$-B\tan\alpha = \tfrac{1}{24}wl^3 + \tfrac{1}{2}M_A l - \tfrac{1}{6}S_{A'}l^2 \quad\dots\dots\dots(4),$$

and by substituting this value in (2) and again putting $x = l$, we find for the value of dy/dx at B

$$B\frac{dy}{dx} = \tfrac{1}{8}wl^3 + \tfrac{1}{2}M_A l - \tfrac{1}{3}S_{A'}l^2;$$

and substituting for $S_{A'}$ from (1), this gives

$$B\frac{dy}{dx} = -\tfrac{1}{24}wl^3 + \tfrac{1}{6}M_A l + \tfrac{1}{3}M_B l \quad\dots\dots\dots\dots(5)$$

for the value of dy/dx at B.

In like manner by taking an origin at B and writing down the bending moment at any point of BC, we get instead of (4) the equation

$$-B\tan\beta = \tfrac{1}{24}wl'^3 + \tfrac{1}{2}M_B l' - \tfrac{1}{6}S_{B'}l'^2,$$

where $\tan\beta$ is the value of dy/dx at B.

Substituting in this the value of $S_{B'}$ from (1), we get

$$B\tan\beta = -\tfrac{1}{24}wl'^3 - \tfrac{1}{2}M_B l' - \tfrac{1}{6}(l'M_C - l'M_B - \tfrac{1}{2}wl'^3)$$
$$= \tfrac{1}{24}wl'^3 - \tfrac{1}{3}M_B l' - \tfrac{1}{6}M_C l' \quad\dots\dots\dots\dots\dots(6).$$

But $\tan\beta$ is the dy/dx of (5), so that, by equating the two values, we get

$$l\,(M_A + 2M_B) + l'\,(2M_B + M_C) = \tfrac{1}{4}w\,(l^3 + l'^3) \quad\dots(7),$$

which is known as 'the equation of three moments'.

13·5. Combined Extension and Bending. The case of combined extension and bending may be illustrated by the following example:

A uniform beam is held inclined at any angle to the vertical in collinear clamps which are at a distance apart equal to the natural length of the beam. Prove that at every inclination the forces on the clamps are equal and vertical, and calculate the couples. [T.]

Let AB be the beam clamped at A and B, where $AB (=l)$ is inclined at an angle α to the vertical. Let w be the weight of unit length of the unstrained beam. Let X_1, Y_1 and X_2, Y_2 be the components of force along and perpendicular to the beam at A and B and M_1, M_2 the couples.

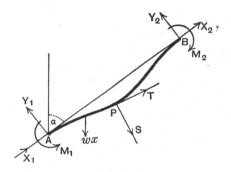

The weight of the beam will cause a variable extension as well as bending. Let $x + \xi$ be the projection on AB of a portion AP whose unstrained length is x, and let y be the distance of P from AB.

Then, if T and S denote the tension and shearing force at P, by resolving for the equilibrium of the portion AP whose weight is wx, we have

$$-S + Y_1 = wx \sin \alpha \quad \dots\dots\dots\dots\dots\dots(1),$$

and

$$T + X_1 = wx \cos \alpha \quad \dots\dots\dots\dots\dots\dots(2).$$

But by **13·21** if ω denotes the cross section of the beam and E is Young's modulus,

$$T = E\omega \frac{d\xi}{dx},$$

therefore from (2)

$$E\omega \frac{d\xi}{dx} + X_1 = wx \cos \alpha;$$

and by integration

$$E\omega\xi + X_1 x = \tfrac{1}{2}wx^2 \cos \alpha,$$

with no constant of integration since ξ and x vanish together. But ξ is also zero when $x = l$, so that

$$X_1 = \tfrac{1}{2}wl \cos \alpha.$$

Also, by resolving along AB for the whole rod, we have
$$X_1 + X_2 = wl \cos \alpha,$$
therefore
$$X_2 = X_1 = \tfrac{1}{2} wl \cos \alpha \quad \dots \dots \dots \dots \dots \dots (3).$$

Again, the bending moment at P is $M = B \dfrac{d^2 y}{dx^2}$.

But
$$-S = \frac{dM}{dx} \ (\mathbf{13·44}),$$
therefore (1) gives
$$B \frac{d^3 y}{dx^3} + Y_1 = wx \sin \alpha,$$
so that, by integration

$$B \frac{d^2 y}{dx^2} + Y_1 x = \tfrac{1}{2} wx^2 \sin \alpha + M_1 \dots \dots \dots \dots (4),$$

M_1 being the value of $B \dfrac{d^2 y}{dx^2}$ at $x = 0$;

therefore
$$B \frac{dy}{dx} + \tfrac{1}{2} Y_1 x^2 = \tfrac{1}{6} wx^3 \sin \alpha + M_1 x \quad \dots \dots \dots \dots (5),$$
and
$$By + \tfrac{1}{6} Y_1 x^3 = \tfrac{1}{24} wx^4 \sin \alpha + \tfrac{1}{2} M_1 x^2 \quad \dots \dots \dots \dots (6),$$

with no other constants of integration since y and dy/dx both vanish with x; but they also vanish for $x = l$ so that from (5) and (6)
$$\tfrac{1}{2} Y_1 l = \tfrac{1}{6} wl^2 \sin \alpha + M_1,$$
and
$$\tfrac{1}{3} Y_1 l = \tfrac{1}{12} wl^2 \sin \alpha + M_1.$$
Whence we find that
$$Y_1 = \tfrac{1}{2} wl \sin \alpha, \quad \text{and} \quad M_1 = \tfrac{1}{12} wl^2 \sin \alpha,$$
and, by resolving at right angles to AB for the whole rod,
$$Y_1 + Y_2 = wl \sin \alpha,$$
therefore
$$Y_2 = Y_1 = \tfrac{1}{2} wl \sin \alpha \quad \dots \dots \dots \dots \dots \dots (7).$$

It follows from (3) and (7) that the resultant forces at A and B are vertical and each equal to $\tfrac{1}{2} wl$.

The couple M_2 is the value of $B \dfrac{d^2 y}{dx^2}$ at $x = l$, but (4) may be written
$$B \frac{d^2 y}{dx^2} + \tfrac{1}{2} wlx \sin \alpha = \tfrac{1}{2} wx^2 \sin \alpha + M_1,$$
and putting $x = l$, we have
$$M_2 = M_1 = \tfrac{1}{12} wl^2 \sin \alpha.$$

13·6. Elastic Energy.

When a bar or beam undergoes a longitudinal strain due to stretching or bending within the elastic limits of the material of which it is composed, its elastic energy is $\tfrac{1}{2} E \epsilon^2$ per unit volume (**13·23**).

This formula may be applied to determine the elastic energy

of the beam in any of the previous examples. Thus for a bent bar (13·3) $\epsilon = y/R$, and the elastic energy per unit length

$$= \frac{1}{2} \frac{E}{R^2} \iint y^2 \, dx \, dy,$$

integrated over the cross section

$$= \frac{1}{2} \frac{EI}{R^2} = \frac{1}{2} B \left(\frac{d^2 y}{dx^2}\right)^2, \quad (13\text{·}4).$$

For example, in 13·42 (i), taking the values of $\frac{d^2 y}{dx^2}$ which correspond to the two parts AC, CB of the beam, we have for the whole elastic energy

$$\frac{1}{2} \frac{W^2 b^2}{Bl^2} \int_0^a x^2 \, dx + \frac{1}{2} \frac{W^2 a^2}{Bl^2} \int_a^l (l-x)^2 \, dx, \quad \text{where} \quad b = l - a,$$

which reduces to

$$\frac{1}{6} \frac{W^2 a^2 (l-a)^2}{Bl}.$$

Since from (5) or (6) the sag at the point C is $\frac{1}{3} \frac{Wa^2 b^2}{Bl}$, therefore the elastic energy is one-half of the loss of potential energy due to the descent of the weight W from the level AB.

13·7.   General Equations for a thin rod bent in one plane.

Let PQ be an element δs of the rod AB in its equilibrium position. Let T, S and M be the tension, shearing force and bending moment at P, and $T + \delta T$, $S + \delta S$ and $M + \delta M$ the corresponding quantities at Q.

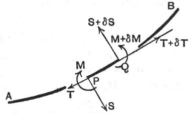

Let $F \delta s$ and $G \delta s$ be the tangential and normal components of the externally applied forces on the element δs. Let $\delta \psi$ be the angle between the tangents at P and Q, so measured that s and ψ increase together.

Then by resolving along the tangent and inward normal at P we have

$$-T + (T + \delta T) \cos \delta \psi - (S + \delta S) \sin \delta \psi + F \delta s = 0,$$

and

$$-S + (S + \delta S) \cos \delta \psi + (T + \delta T) \sin \delta \psi + G \delta s = 0.$$

In the limit as Q approaches P these become

$$\frac{dT}{ds} - \frac{S}{\rho} + F = 0 \quad \dotsfill (1),$$

and

$$\frac{dS}{ds} + \frac{T}{\rho} + G = 0 \quad \dotsfill (2),$$

where ρ is the radius of curvature at P.

Also, by taking moments about Q,

$$-M + (M + \delta M) + S\,\delta s - G\,\delta s\,(\theta\,\delta s) = 0, \qquad 0 < \theta < 1,$$

or

$$\frac{dM}{ds} + S = 0 \quad \dotsfill (3).$$

Since the strained form of the rod is not known, another equation is necessary to determine the form in addition to T, S and M. This equation is

$$M = B\left(\frac{1}{\rho} - \frac{1}{\rho_0}\right) \quad \dotsfill (4),$$

where B is the flexural rigidity and ρ_0 is the radius of curvature at P before the deformation. On the hypotheses of **13·3**, this equation can be established by a similar proof, the extension ϵ of an element at a distance y from the neutral line in this case being $y\left(\dfrac{1}{\rho} - \dfrac{1}{\rho_0}\right)$.

13·8. Euler's Strut.* Consider the possible positions of equilibrium of a long thin rod clamped vertically at its lower end and carrying a weight W at its upper end, on the hypothesis of a deflection of the upper end from the vertical.

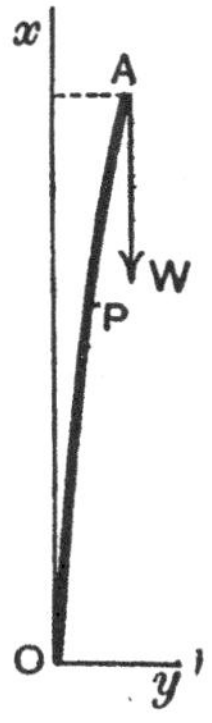

Let the rod be of length l and suppose that the upper end is at a distance a from the vertical through the lower end. Then, taking axes as shown in the figure, for the bending moment at a point $P(x, y)$ we have

$$B\frac{d^2 y}{dx^2} = W(a - y).$$

The primitive of this equation is

$$y = a + F\cos\sqrt{(W/B)}\,x + G\sin\sqrt{(W/B)}\,x,$$

* L. Euler, 1757.

and if we determine the constants F, G so as to make $y = 0$ when $x = 0$ and $y = a$ when $x = l$, the solution becomes

$$y = a \left\{ 1 - \frac{\sin \{\sqrt{(W/B)}\,(l-x)\}}{\sin \{\sqrt{(W/B)}\,l\}} \right\}.$$

But dy/dx is to vanish for $x = 0$, and this requires that $\cos \{\sqrt{(W/B)}\,l\} = 0$, and the smallest value of l for which this is possible is $\frac{1}{2}\pi\sqrt{(B/W)}$.

The inference to be drawn from this result is that if the rod were shorter than this critical length $\frac{1}{2}\pi\sqrt{(B/W)}$ it would remain vertical and merely contract under the load W, but if the length were increased up to the critical length an equilibrium form is that of a portion of a sine curve of small amplitude, and it can be shewn by calculating the potential energy that the vertical position is stable when the length is less than the critical length and that when the critical length is slightly exceeded the stability passes from the straight to the bent form.

13·81. The problem of a long thin rod AB (fig. (i)) of length l whose ends are subject to equal and opposite forces W is seen to be subject to the same analysis as the problem of **13·8**, if we place an origin O at the middle point of the rod, and in the result of **13·8** substitute $\frac{1}{2}l$ for l, and the rod will bend under the given forces if

$$l^2 W > \pi^2 B.$$

Similarly, if the ends C, D are both clamped in the same straight line (fig. (ii)) and the rod be of length l and subject to the same terminal forces W and we draw a sine curve touching CD at C and D and of small amplitude, the distance between its inflexions A, B is $\frac{1}{2}l$ and the problem for the portion AOB of the rod is the same as for the rod AOB of fig. (i), so that the condition for bending in this case is

$$l^2 W > 4\pi^2 B.$$

(i)   (ii)

13·9. Loaded Column. When a vertical column supports a load the stress at every point of a cross section of the column ought to be of the same sign—for, if the stress were a compression over part of the cross section and a tension over the remaining part, the result might be a fracture of the column.

In **13·3** fig. (ii) EOF is the neutral line of a cross section of a bar and the parts of the bar above the neutral line are extended and in a state of tension while the parts below are compressed and in a state of thrust; and it is clear that in order that the stress may have the same sign at all points of a cross section the neutral line must fall outside the cross section.

The position of the neutral line depends on the state of stress and this depends on the line of action of the resultant load.

Now the stress across an element $dx\,dy$ of the cross section is $\dfrac{Ey}{R}\,dx\,dy$

(13·3), giving a resultant $\dfrac{E}{R}\displaystyle\iint y\,dx\,dy=\dfrac{E\omega}{R}\,\bar{y}$, where ω is the area of the cross section and $\bar{y}$ is the distance of its centroid from the neutral line. And the line of action of the resultant stress is at a distance η from the neutral line obtained by taking moments about the neutral line, thus

$$\eta\,\frac{E\omega}{R}\,\bar{y}=\frac{E}{R}\iint y^2\,dx\,dy \quad\text{or}\quad \eta=\frac{k^2}{\bar{y}},$$

where k is the radius of gyration of the cross section about the neutral line.

Now let Ox be the neutral line when the load W acts in the line LC cutting the cross section in C; and let G be the centroid of the cross section.

Then W is equal to the resultant stress $\dfrac{E\omega}{R}\,\bar{y}$, OG is $\bar{y}$ and OC is η; and if the shape of the cross section is known, the data are now sufficient to determine within what area the point C must lie in order that Ox may fall outside the cross section.

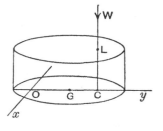

Thus if the section is a circle of radius a, by the theorem of parallel axes,

$$k^2=\bar{y}^2+\frac{a^2}{4};$$

and Ox will fall outside the circle if $\bar{y}>a$. But

$$OC=\eta=\frac{k^2}{\bar{y}}=\bar{y}+\frac{a^2}{4\bar{y}}, \quad\text{so that}\quad GC=\frac{a^2}{4\bar{y}},$$

and therefore we shall have $\bar{y}>a$ if $GC<\tfrac{1}{4}a$.

This means that in order that the stress in a circular column may nowhere be a tension, the line of action of the load must cut the cross section within a concentric circle of radius $\tfrac{1}{4}a$.

EXAMPLES

1. A uniform beam of length $6a$ rests symmetrically upon two supports distant $2a$ apart. Find the elevation of the highest point above the horizontal line joining the supports.   [I.]

2. A uniform horizontal beam, of length l and negligible weight, is supported at the ends, and carries a weight W at the middle

point. Find the deflection at this point, and prove that the elastic energy is $\dfrac{W^2 l^3}{96 EI}$. [I.]

3. Prove that the elastic energy of a uniform beam of weight W and length l supported at its ends at the same level is $\frac{1}{240} W^2 l^3 / B$.

4. Prove that the elastic energy of a uniform rod of weight W and length l hanging from one end is $\frac{1}{6} W^2 l / E\omega$.

5. Find the bending moment at any point of a uniform heavy beam of length $2a$ supported at its middle point; find also the depression of the ends and shew that it is increased if a weight W be suspended from each end by an amount $\dfrac{1}{3}\dfrac{Wa^3}{B}$, where B is the flexural constant. [I.]

6. Prove that the deflection of the middle point of a uniform homogeneous elastic beam of length $2a$ supported at two points distant b from the middle point and lying in the same horizontal line is

$$\frac{1}{24}\frac{w}{B}(6a^2 b^2 - 12ab^3 + b^4),$$

where w is the weight per unit length of the beam and B the flexural rigidity. [I.]

7. A uniform slightly flexible rod AB, of length $2a$, is supported at its two ends, and also at its middle point M; the end supports are in a horizontal line and the middle support at a small depth δ below this line. Shew that the reaction at M is decreased by an amount $6EI\delta/a^3$ as compared with its value if AMB were horizontal. [I.]

8. Prove that, if a heavy uniform beam of length l is supported at two points at the same level, at one end and at a distance $\frac{3}{4}l$ from that end, then the beam is horizontal at the latter point.

9. A slightly flexible rod of length $2a$ has one end clamped horizontally; a support is placed under the middle point of the rod so that the free end is in the same horizontal line as the fixed end. Prove that the height of the middle point above the ends is $\dfrac{11Wa^3}{240B}$, where B is the flexural constant and W is the weight of the rod.

Shew also that the pressure on the support is $\frac{5}{8}W$. [I.]

10. A uniform thin beam of length l is clamped at one extremity A so that the tangent at it is horizontal; and the other extremity B rests on a fixed support in the same horizontal line with A. Shew that there is a point of inflexion at a distance $\frac{1}{4}l$ from A, and that the tangent is horizontal at a distance

$$(15-\sqrt{33})\,l/16$$

from A; and that the pressure on B is $\frac{3}{8}$ the total weight of the beam. [I.]

11. If a rod be clamped horizontally at each end at the same level and the middle point be pulled upwards by a force through a distance δ above the level of the ends, prove that the magnitude of the force is $\dfrac{24B\delta}{a^3} + \dfrac{W}{2}$ and that the bending couples at the ends are equal to $\dfrac{6B\delta}{a^2} - \tfrac{1}{24}Wa$, where $2a$ is the length of the rod, W its weight, and B the flexural rigidity. [I.]

12. Shew that, with the notation of 13·42 (i), if the ends of the beam AB are clamped horizontally, the supporting forces at A and B are
$$W(3a+b)\,b^2/l^3 \quad \text{and} \quad W(a+3b)\,a^2/l^3,$$
and the bending moments at A and B are
$$Wab^2/l^2 \quad \text{and} \quad Wa^2b/l^2.$$

13. A light beam is supported at its ends at the same level. Prove that the deflection at a point P when there is a load at Q is equal to the deflection at Q when there is an equal load at P.

14. Prove the same theorem when the ends of the beam are clamped in the same horizontal line.

15. A uniform beam rests with its ends and middle point supported upon three rigid props at the same level. Prove that the pressure on the middle prop is $\tfrac{5}{8}$ of the weight of the beam.
Prove further, that if the weight of the beam, instead of being uniformly distributed, is concentrated in two equal weights at the middle points of the two spans, the pressure on the middle prop is $\tfrac{11}{16}$ of the weight. [T.]

16. A thin uniform elastic beam of length l, which is straight when unstrained, is supported at two points which are at the same level and are equidistant from the centre of the beam. How far apart must the supports be placed in order that the height of the middle of the beam above this level may be as great as possible? [T.]

17. A uniform beam of length $2l$ is supported at the ends: the deflection at the centre is y_1 when a load rests on the centre, and is y_2 when the same load rests on the middle of another shorter beam (of length $2a$) whose ends are supported by the former at points equidistant from the middle. Prove that
$$\frac{y_2}{y_1} = 1 - \frac{3}{2}\frac{a^2}{l^2} + \frac{1}{2}\frac{a^3}{l^3}. \tag{T.}$$

18. A light beam PQ, of span $2a$, rests on two supports at P and Q in the same horizontal line. It is loaded so that the weight of the load per unit length at any point X varies as the square of PX. Prove that the deflection at the middle point of the beam is $89wa^4/(1440B)$, where B is the flexural rigidity of the beam and w is the weight of the load per unit length at Q. [T.]

19. A uniform beam $ABC\,(AB=BC)$ is supported at A, B, C so that A, B, C are horizontal and is loaded at D, E, the midpoints of AB, BC respectively, with loads in the ratio $1:3$. The weight of the beam is neglected. Prove that the beam is horizontal at A. [T.]

20. A spring-board consists of a heavy uniform beam of length $a+b$ clamped horizontally at one end and with a support of small height h at a distance a from this end. If a weight W is placed on the free end, shew that the difference of level of that end before and after the weight is placed upon it is $Wb^2\,(3a+4b)/12B$, where B is the flexural rigidity of the beam. [T.]

21. AD is a heavy, uniform beam resting on supports at its ends and at points B, C in the horizontal line AD; $AB=2a$, $BC=a$, $CD=a$. Find the ratios of the pressures on the four supports. [I.]

22. Shew that, if a thin rod of length l and flexural rigidity B is set up vertically with its lowest point fixed and the rod passing through a smooth ring fixed at a vertical height $\frac{1}{2}l$ above the lowest point and carrying at the top a weight W, then the least value of l for which the rod bends under the load is the smallest root of the equation $\tan\frac{1}{2}nl=nl$, where
$$n^2=W/B.\qquad\text{[T.]}$$

23. A light uniform bar of length l is placed in a vertical position, the lower end being clamped and the upper end free. A force, whose horizontal and downward vertical components are F and W, is applied at the centroid of the cross section of the upper end so as to cause bending in the plane containing the axis of symmetry of a cross section and the axis of the bar, W being smaller than the vertical force for which the vertical position of the bar is unstable. Assuming the inclination of the bar to the vertical to be everywhere small, shew that the deflection of the free end is $\dfrac{Fl}{W}\left(\dfrac{\tan nl}{nl}-1\right)$, where $n^2=W/B$ and B is the flexural rigidity.

If the cross section of the bar is a circle of radius a, shew that, in order that the normal stress across a cross section should nowhere be a tension,
$$W\frac{nl}{\tan nl}\geqslant\frac{4Fl}{A}.\qquad\text{[T.]}$$

ANSWERS

w denotes weight per unit length

1. $19wa^4/24B$.   2. $Wl^3/48B$.   5. $wa^4/8B$.
16. $\frac{1}{2}(9-\sqrt{69})\,l$.   21. $75:186:64:43$.

Chapter XIV

FORCES IN THREE DIMENSIONS

14·1. *A system of forces acting on a rigid body can be reduced in general to a force acting at a specified point of the body and a couple.*

Take a set of rectangular axes with the specified point O as origin and let x, y, z be the co-ordinates of the point of application A of a typical force P, and let the components of P parallel

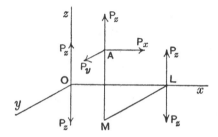

to the axes be P_x, P_y, P_z. Let M be the projection of A on the plane xy and L its projection on Ox. At L and at O introduce pairs of equal and opposite forces equal and parallel to P_z. These do not affect the resultant of the system. There are now five forces of magnitude P_z and they constitute a force P_z along Oz, a couple of forces P_z with an arm ML, i.e. a couple of moment yP_z in or parallel to the plane yOz in the sense from Oy to Oz, and a couple of forces P_z with an arm OL, i.e. a couple of moment xP_z in the plane zOx in the sense from Ox to Oz.

Hence, adopting the convention of signs for couples explained in **4·641**, it appears that the force P_z acting at A is equivalent to an equal and parallel force at O together with a couple whose axis is yP_z along Ox and a couple whose axis is $-xP_z$ along Oy.

By similar steps the forces P_x, P_y acting at A can be transferred to act along Ox and Oy, the necessary couples being

$$zP_x \text{ along } Oy \text{ and } -yP_x \text{ along } Oz,$$

and xP_y along Oz and $-zP_y$ along Ox.

Hence the force P acting at A is equivalent to component forces P_x, P_y, P_z acting along the axes at O, and couples of moment $yP_z - zP_y$ with axis along Ox,

$$zP_x - xP_z \text{ with axis along } Oy,$$

$$xP_y - yP_x \text{ with axis along } Oz.$$

All the forces of the given system can be treated in this way, so that the given system is equivalent to a force R acting at O whose components are

$$X = \Sigma P_x, \quad Y = \Sigma P_y, \quad Z = \Sigma P_z;$$

together with a couple G whose components are

$$L = \Sigma(yP_z - zP_y), \quad M = \Sigma(zP_x - xP_z), \quad N = \Sigma(xP_y - yP_x).$$

14·11. It is clear that X, Y, Z are the algebraical sums of the resolved parts of the given forces in the directions of the co-ordinate axes, and by reference to **4·21** it is also clear that L, M, N are the algebraical sums of the moments of the given forces about the co-ordinate axes.

Also R is the resultant of the given forces moved parallel to themselves to act at the point O and is clearly independent of the position of the point O. Hence R^2 or $X^2 + Y^2 + Z^2$ is said to be *an invariant* of the system of forces.

The values of G and its components depend on the choice of origin. For if the system were transferred to a point O' whose co-ordinates are ξ, η, ζ referred to the chosen axes through O, the only difference would be in the lengths of the arms of the couples. Instead of x, y, z we should have $x - \xi$, $y - \eta$, $z - \zeta$, so that the components of the resultant couple G' would be L', M', N', where

$$L' = \Sigma\{(y - \eta)P_z - (z - \zeta)P_y\}$$
$$= \Sigma(yP_z - zP_y) - \eta\Sigma P_z + \zeta\Sigma P_y$$
$$= L - \eta Z + \zeta Y,$$

and similarly $M' = M - \zeta X + \xi Z,$

and $N' = N - \xi Y + \eta X.$

It follows that
$$L'X + M'Y + N'Z = LX + MY + NZ.$$

14·2. Conditions of Equilibrium. It is necessary for equilibrium that there shall be neither resultant force nor resultant couple; i.e. that $R = 0$ and $G = 0$

or that $X^2 + Y^2 + Z^2 = 0$ and $L^2 + M^2 + N^2 = 0.$

These conditions can only be satisfied if
$$X = 0, \quad Y = 0, \quad Z = 0 \quad \text{and} \quad L = 0, \quad M = 0, \quad N = 0.$$

Since the choice of axes is arbitrary, this means that the algebraical sum of the resolved parts of the forces in every direction must be zero and the algebraical sum of the moments of the forces about every straight line must be zero.

To ensure the vanishing of R it will be sufficient to know that the sums of the resolved parts of the forces in any three directions not all parallel to the same plane are zero. For if OA, OB, OC are lines in the given directions, the fact that R has no component in direction OA means that R is either zero or at right angles to OA, and it cannot be at right angles to each of three non-coplanar directions, therefore it is zero.

Similarly, since couples are compounded by the vector law, to ensure the vanishing of G it is sufficient to know that the algebraical sums of the moments of the forces about any three lines meeting in O and not coplanar are zero.

14·21. *Special cases.*

(i) *When R is zero and G is not zero.* The system reduces to a couple G. The vector which represents the axis of this couple may be placed at any point of space. The algebraical sum of the moments of the forces about any line parallel to the axis of the couple is G, and the algebraical sum of the moments of the forces about a line inclined at an angle θ to the axis of the couple is $G \cos \theta$.

(2) *When G is zero.* The system reduces to a single resultant force R.

14·3. Poinsot's Central Axis. The Wrench. When neither R nor G is zero.

Let the axis of the couple G make an angle θ with the direction of R at the point O as in fig. (i). G can then be resolved into a couple $G\cos\theta$ with axis along R and $G\sin\theta$ with axis perpendicular to R as in fig. (ii), in which the vector $G\sin\theta$ is supposed to be at right angles to the plane of the paper. The couple $G\sin\theta$ is then represented by two forces of magnitude R at a distance $G\sin\theta/R$ apart, so placed that one of them balances the given force R at O and the other acts in a parallel line through O', as in fig. (iii). Removing the equal and opposite forces at O, there remain (fig. (iv)) the force R at O'

(i)   (ii)   (iii)   (iv)

and a couple $G\cos\theta$ whose axis is in the direction of R, so that the system has been reduced to a single force R at O' and a couple whose axis coincides in direction with the force, or a force along a certain line and a couple in a plane perpendicular to the line. Such a combination of force and couple is called a **wrench**. The line of action of the force is called **Poinsot's central axis** after L. Poinsot who first investigated the theory*. It is also called the *axis of the wrench*; the *pitch* of the wrench is defined to be the ratio of the moment of the couple to the magnitude of the force, and the magnitude of the force is called the *intensity of the wrench*.

14·31. *Equations of the Central Axis.* With the notation of **14·11** let (ξ, η, ζ) be a point on the central axis. The system reduces to a force R at this point and a couple whose magnitude we will denote by Γ, whose axis coincides in direction with R.

---

* *Éléments de Statique,* 1804.

But the direction cosines of R are X/R, Y/R, Z/R and those of Γ are L'/Γ, M'/Γ, N'/Γ, so that

$$\frac{L'}{X} = \frac{M'}{Y} = \frac{N'}{Z} = \frac{\Gamma}{R},$$

or $\quad \dfrac{L - \eta Z + \zeta Y}{X} = \dfrac{M - \zeta X + \xi Z}{Y} = \dfrac{N - \xi Y + \eta X}{Z} = \dfrac{\Gamma}{R} = p$

$$\ldots\ldots(1),$$

where p is the pitch of the resultant wrench.

The equations with ξ, η, ζ as current co-ordinates are the equations of the central axis.

By whatever process we reduce the forces to the equivalent wrench we must arrive at the same force R and couple Γ. In 14·3 the couple of the wrench is $G \cos\theta$; therefore

$$G \cos\theta = \Gamma.$$

But θ is the angle between R, whose direction cosines are as above, and G, whose direction cosines are L/G, M/G, N/G, so that

$$\cos\theta = \frac{LX + MY + NZ}{RG}.$$

Therefore $\quad LX + MY + NZ = R\Gamma \quad \ldots\ldots\ldots\ldots\ldots(2)$

is an invariant of the system of forces.

The pitch p of the resultant wrench may also be expressed in terms of the component forces and couples, thus

$$p = \frac{LX + MY + NZ}{X^2 + Y^2 + Z^2}.$$

14·32. As stated in **14·31** there can only be one central axis for a given system of forces. For if there were two equivalent wrenches with different axes, then one reversed would balance the other. But the forces R are the same for both wrenches and when one is reversed they would constitute a couple balancing a couple in a perpendicular plane, which is not possible.

14·33. *Condition for a Single Resultant Force.* When Γ is zero and R is not zero, the system has a single resultant R. The condition for this is by **14·31** (2)

$$LX + MY + NZ = 0 \quad \ldots\ldots\ldots\ldots\ldots(1).$$

In this case, at any point (ξ, η, ζ) on the line of action of the single resultant, the components of couple L', M', N' vanish:

i.e.
$$L - \eta Z + \zeta Y = 0,$$
$$M - \zeta X + \xi Z = 0,$$
and
$$N - \xi Y + \eta X = 0.$$

In virtue of (1) these three equations are not independent, but any two of them may be regarded as the equations of the line of action of the single resultant force.

14·34. *Given the resultant wrench to find the force and couple at any point.*

Let R, Γ be the force and couple of the wrench and a line Oz its axis —fig. (i).

To find the force and couple at a point O'; let $O'O = h$ be perpendicular to Oz. At O' introduce opposite forces equal and parallel to R—fig. (ii).

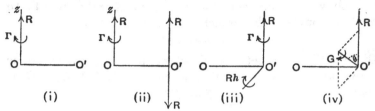

(i)   (ii)   (iii)   (iv)

One of these constitutes, with R at O, a couple of moment Rh, whose axis is at right angles to the plane $O'Oz$; and the axis of the couple Γ can be placed at O'—fig. (iii). The two couples Γ and Rh can then be compounded into a single couple G whose axis makes an angle θ with R, where
$$G^2 = \Gamma^2 + R^2 h^2 \quad \text{and} \quad \tan\theta = Rh/\Gamma,$$
and thus the wrench in the line Oz is equivalent to the force R and couple G at O'.

14·341. Example. *Forces P, Q, R, P, Q, R act along the edges BC, CA, AB, AD, BD, CD of a regular tetrahedron $ABCD$. Shew that they are equivalent to a wrench of pitch $a/2\sqrt{2}$, where a is the length of an edge.* [I.]

It is convenient to take as the six edges of the tetrahedron six diagonals of the faces of a cube as in the figure. Then, taking axes at the centre of the cube parallel to its edges, we have
$$X = P\sqrt{2}, \quad Y = R\sqrt{2}, \quad Z = Q\sqrt{2},$$
and
$$L = -\frac{2P}{\sqrt{2}} \cdot \frac{a}{2\sqrt{2}}, \quad M = -\frac{2R}{\sqrt{2}} \cdot \frac{a}{2\sqrt{2}}, \quad N = -\frac{2Q}{\sqrt{2}} \cdot \frac{a}{2\sqrt{2}};$$
for the length of the edge of the cube is $a/\sqrt{2}$.

Then the pitch being numerically equal to

$$(LX + MY + NZ)/(X^2 + Y^2 + Z^2)$$

is $a/2\sqrt{2}.$

14·4. The Invariants. The invariants of a system of forces are the resultant force R, and the expression $LX + MY + NZ$ which is equal to the product of the force and couple of the resultant wrench and will be denoted by I.

Since R is the resultant of all the forces moved parallel to themselves to act at a point, it follows from **3·31** that for any system of forces P_1, P_2, P_3, ...

$$R^2 = \Sigma P_r{}^2 + 2\Sigma P_r P_s \cos\theta_{rs},$$

where θ_{rs} denotes the angle between the forces P_r, P_s and the latter sum includes the products of all the forces taken in pairs.

To express I in terms of the forces, their inclinations and their distances apart:

(i) Consider first the case of *two forces P_1, P_2 inclined at an angle θ at a distance h apart*. Take the axis Oz along P_1 and the axis Ox along the common perpendicular to the lines of action.

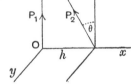

Then

$$X = 0, \quad Y = P_2 \sin\theta, \quad Z = P_1 + P_2 \cos\theta,$$
$$L = 0, \quad M = -P_2 h \cos\theta, \quad N = P_2 h \sin\theta.$$

Therefore

$$I = LX + MY + NZ = P_1 P_2 h \sin\theta.$$

(ii) *A force R and a couple G whose axis makes an angle θ with the direction of R.* Since, by **14·3**, these are equivalent to a wrench whose force is R and couple $G \cos\theta$, therefore in this case

$$I = RG \cos\theta.$$

(iii) *Two couples.* The couples can be so placed that all the four forces are parallel and then by (i), since $\theta = 0$, I is zero for each pair of the forces and therefore for the couples.

(iv) *General case.* By definition I is a sum of products of forces and couples and is therefore of two dimensions in force. We may therefore assume that I is a quadratic expression in terms of the forces and write

$$I = a_{11} P_1{}^2 + a_{22} P_2{}^2 + \ldots + 2a_{12} P_1 P_2 + \ldots,$$

where the coefficients do not depend upon the magnitudes of the forces but only on their relative positions.

To find the coefficients let us suppose that all the forces are zero except P_1 and P_2, then from (i) we have

$$I = P_1 P_2 h_{12} \sin\theta_{12},$$

where h_{12} denotes the shortest distance and θ_{12} the angle between P_1 and P_2.

By comparison it follows that

$$a_{11} = a_{22} = 0,$$

and

$$2a_{12} = h_{12} \sin \theta_{12}.$$

Similarly the other coefficients are determined and we have, for any number of forces, $I = \Sigma P_r P_s h_{rs} \sin \theta_{rs},$

where the summation includes the products of all the forces taken in pairs.

It is evident that in this sum there may be terms of opposite signs and we need a convention for the determination of the sign of each term.

Each term may be regarded as the product of a force P_r and the moment of a force P_s about the line of action of P_r. We must then choose either a right-handed or a left-handed screw convention (**4·641**) and take the term as positive or negative according as the vector which represents the moment of P_s about P_r is in the same sense as the force P_r or in the opposite sense.

(v) *Two wrenches of intensities* P_1, P_2 *and pitches* p_1, p_2 *at a distance* h *apart and inclined at an angle* θ. There are two forces P_1, P_2 and two couples $p_1 P_1$, $p_2 P_2$, and we have the following contributions:

from the two forces	$P_1 P_2 h \sin \theta$	by (i)
from the two couples	0	by (ii)
from the first force and couple	$p_1 P_1^2$	...
from the second force and couple	$p_2 P_2^2$	...
from the first force and second couple	$P_1 . p_2 P_2 \cos \theta$	...
from the second force and first couple	$P_2 . p_1 P_1 \cos \theta$	...

Therefore $I = p_1 P_1^2 + p_2 P_2^2 + P_1 P_2 \{(p_1 + p_2) \cos \theta + h \sin \theta\}.$

14·41. Geometrical Representation of the Moment of a Force about a Line.

Lemma. The volume of a tetrahedron is one-sixth of the product of a pair of opposite edges, the shortest distance between them and the sine of their inclination.

Let $DABC$ be the tetrahedron. Complete the parallelogram $ABEC$, and by taking parallel planes complete the parallelepiped as shewn in the figure.

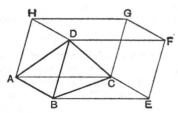

Then the volume $DABC$

$= \frac{1}{3}$ base × height

$= \frac{1}{6}$ volume parallelepiped

$= \frac{1}{6}$ area $BEFD$ × distance between $BEFD$ and $ACGH$

$= \frac{1}{6} BD . BE \sin DBE$ × shortest distance between BD and AC

$= \frac{1}{6} BD . AC$ × shortest distance between BD and AC × $\sin (\widehat{BD.AC})$.

Now let P be a force in a line AB, then by definition its moment about a line CD is $Ph\sin\theta$, where h is the shortest distance and θ the angle between the lines.

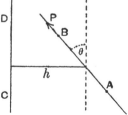

But, if V is the volume of the tetrahedron $ABCD$,

$$V = \tfrac{1}{6}AB\,.\,CD\,.\,h\sin\theta;$$

therefore the moment of P about CD

$$= \frac{6V\,.\,P}{AB\,.\,CD}.$$

The choice of the lengths AB, CD is arbitrary. If we like to take them of unit length, then the moment of P about CD will be measured by $6P$ times the volume of $ABCD$.

Further, if two forces P_1, P_2 are represented by straight lines AB, CD, then by **14·4** (i) the invariant I of the forces is

$$P_1 P_2 h\sin\theta, \quad \text{or} \quad AB\,.\,CD h\sin\theta,$$

i.e. six times the volume of the tetrahedron $ABCD$, with due regard to sign as explained in **14·4** (iv).

14·42. Example. *If $ABCD$ be a tetrahedron and forces are represented by pBC, qCA, rAB, $p'DA$, $q'DB$, $r'DC$, shew that they are equivalent to a single force if $pp' + qq' + rr' = 0$.* [I.]

By **14·33** the condition for a single resultant force is

$$LX + MY + NZ = 0,$$

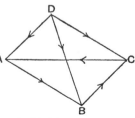

or the vanishing of the invariant I. But by **14·4** (i) two forces which intersect contribute nothing to I, so that the only contributions in this case arise from the forces in pairs of opposite sides of the tetrahedron; i.e. from pBC and $p'DA$, from qCA and $q'DB$, and from rAB and $r'DC$.

And from **14·41** the contribution of pBC and $p'DA$ is

$$6pp' \times \text{volume } DABC,$$

and similarly for the other pairs, so that the invariant I vanishes if

$$pp' + qq' + rr' = 0.$$

14·5. *To find a wrench equivalent to two forces P, P' inclined at an angle θ at a distance h apart.*

(i) The intensity R of the wrench and the pitch ϖ may be written down at once by equating invariants. Thus

$$R^2 = P^2 + P'^2 + 2PP'\cos\theta,$$

and

$$R^2\varpi = PP'h\sin\theta.$$

(ii) Or we may proceed thus:

Let the shortest distance between the forces be AA'. Determine an angle α such that $\qquad P\sin\alpha = P'\sin(\theta-\alpha)$ ....(1).

Then parallel lines Az, $A'z'$ may be drawn making angles α and $\theta-\alpha$ with P and P'. Resolve P into $P\cos\alpha$ and $P\sin\alpha$, and P' into $P'\cos(\theta-\alpha)$ and $P'\sin(\theta-\alpha)$. Two of these forces $P\cos\alpha$, $P'\cos(\theta-\alpha)$ act in parallel lines Az, $A'z'$ and the other two are by (1) equal and opposite parallel forces in a plane perpendicular to Az and $A'z'$.

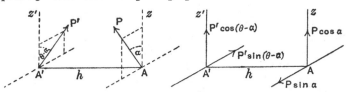

Also since $\qquad P\sin\alpha = P'\sin(\theta-\alpha)$,

therefore $\dfrac{\sin\alpha}{P'\sin\theta} = \dfrac{\cos\alpha}{P+P'\cos\theta} = \dfrac{1}{\sqrt{(P^2+P'^2+2PP'\cos\theta)}} = \dfrac{1}{R}\ldots(2)$,

and the moment of the couple is $Ph\sin\alpha$ or $PP'h\sin\theta/R$ as above.

The position of the wrench is found from the fact that R is the resultant of the parallel forces $P\cos\alpha$ and $P'\cos(\theta-\alpha)$, and therefore the axis of the wrench divides AA' in the ratio

$\qquad\qquad P'\cos(\theta-\alpha) : P\cos\alpha,$

or $\qquad\qquad P'(\cos\theta\cos\alpha + \sin\theta\sin\alpha) : P\cos\alpha,$

i.e. $\qquad P'\cos\theta(P+P'\cos\theta) + P'^2\sin^2\theta : P(P+P'\cos\theta),$

or $\qquad\qquad P'(P'+P\cos\theta) : P(P+P'\cos\theta).$

14·51. *To find the resultant of two given wrenches* (P, p), (P', p') *at a distance h apart and inclined at an angle θ.*

By equating invariants we get for the intensity R and pitch ϖ of the equivalent wrench $\qquad R^2 = P^2 + P'^2 + 2PP'\cos\theta,$

and $\qquad R^2\varpi = pP^2 + p'P'^2 + PP'\{(p+p')\cos\theta + h\sin\theta\}.$

Further, if AA' be the shortest distance between the axes of the given wrenches neither
wrench has any moment about AA', so that
the resultant wrench can have no moment
about AA' and its axis must therefore intersect
AA' at right angles.

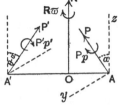

The direction of R is determined by the angle
α of **14·5** (1), and its distance from A say is
obtained by equating moments about a line Ay
perpendicular to AA' and Az. Thus

$\qquad R.OA = Pp\sin\alpha - P'p'\sin(\theta-\alpha) + P'h\cos(\theta-\alpha),$

which by **14·5** (2) reduces to

$\qquad R^2.OA = PP'(p-p')\sin\theta + P'(P\cos\theta+P')h.$

14·6. Nul Points, Lines and Planes. In relation to a given system of forces a *nul line* is a line about which the sum of the moments of the forces is zero. Since a system of forces can be reduced to a single force R acting at a chosen point O together with a couple G, therefore a line through O is a nul line if, and only if, it lies in the plane of the couple G. This plane is called the *nul plane* of the point O and O is called the *nul point* of the plane. In other words the nul plane of the point O is the plane of the principal couple G at the point O.

14·61. Consider a line in the nul plane of O but not passing through O. If it were a nul line the moment of R about it would be zero, which requires either that R is zero or that R itself lies in the nul plane. In the one case the system is equivalent to a couple and in the other to a single force in the nul plane of the point O. Also when R lies in the nul plane of the point O every line in the plane is a nul line and every point of the plane is a nul point. And in both the cases included in this paragraph the invariant I of the system is zero.

14·62. From the definitions of **14·6** it is evident that the nul planes of all points on a nul line contain the line, so that in general as a point travels along a nul line the nul plane of the point turns round the line.

It follows that *if the nul plane of a point A passes through B then the nul plane of B passes through A*, for the line AB is a nul line in both the nul planes.

14·63. *The equation of the nul plane of a point* $(\xi,\ \eta,\ \zeta)$. In the notation of **14·11** it is the plane of the couple G', and the direction cosines of the normal to the plane are proportional to $L',\ M',\ N'$. Therefore the equation of the plane is

$$L'(x-\xi)+M'(y-\eta)+N'(z-\zeta)=0,$$

or, giving $L',\ M',\ N'$ their values,

$$(L-\eta Z+\zeta Y)(x-\xi)+(M-\zeta X+\xi Z)(y-\eta)$$
$$+(N-\xi Y+\eta X)(z-\zeta)=0,$$

which reduces to

$$(L-\eta Z+\zeta Y)x+(M-\zeta X+\xi Z)y+(N-\xi Y+\eta X)z$$
$$=L\xi+M\eta+N\zeta \quad \dots(1).$$

14·64. *To find the nul point of a given plane.*

Take any two points on the given plane and as in **14·63** write down the equations of their nul planes. These planes intersect the given plane in the point required.

14·641. Example. *Find the nul point of the plane $x + y + z = 0$, for the force system (X, Y, Z, L, M, N).*

The points $(0, 0, 0)$ and $(1, -1, 0)$ both lie on the given plane

$$x + y + z = 0 \quad \dots\dots\dots\dots\dots\dots\dots\dots(1).$$

Using **14·63** (1), the nul plane of $(0, 0, 0)$ is

$$Lx + My + Nz = 0 \quad \dots\dots\dots\dots\dots\dots(2),$$

and the nul plane of $(1, -1, 0)$ is

$$(L + Z) x + (M + Z) y + (N - Y - X) z = L - M \quad \dots\dots(3).$$

From (1) and (2) we have

$$\frac{x}{N - M} = \frac{y}{L - N} = \frac{z}{M - L},$$

and from (3) these also

$$= \frac{L - M}{(N - M)(L + Z) + (L - N)(M + Z) + (M - L)(N - Y - X)}$$

$$= \frac{1}{X + Y + Z}.$$

These relations give the co-ordinates of the required nul point.

14·7. Conjugate Forces. *A system of forces can in general be reduced to two forces one of which acts in a given straight line.*

Let the system be equivalent to a force R acting at any point O in the given straight line OA together with a couple G, fig. (i). The plane through OA and R cuts the plane of the

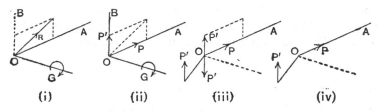

(i)   (ii)   (iii)   (iv)

couple G in another line OB through O. Let R be resolved into oblique components P, P' in the lines OA, OB, fig. (ii). The couple G may then be represented by two equal and opposite parallel forces of magnitude P' so placed that one of them balances the force P' along OB, fig. (iii). The system has then been reduced to a force P along OA and a force P' acting in the nul plane of the point O, fig. (iv). The magnitudes of the forces are connected by the relations

$$\frac{P}{\sin BOR} = \frac{P'}{\sin AOR} = \frac{R}{\sin AOB}.$$

The forces P and P' are called *conjugate forces* and the lines in which they act are called *conjugate lines* with respect to the given system of forces.

Since the conjugate of OA lies in the nul plane of O, therefore as the point O moves along OA its nul plane turns round this conjugate line.

14·71. Exceptional Cases.

(i) $R = 0$. (α) The line OA may be inclined to the plane of the couple G; then the force P in OA is zero and the conjugate force P' is at an infinite distance and also zero. (β) OA may lie in the plane of the couple; the force P is then one force of the couple and its conjugate is the equal and opposite parallel force, the distance between them depending on the magnitude of P which may be chosen arbitrarily.

(ii) *R lies in the nul plane of O.* In this case the system reduces to a single force R. (α) If OA intersects the single force R, then any line through O in the plane of OA and R may be taken as the conjugate line, and R may be resolved along OA and the chosen direction. (β) If OA does not intersect the single force R, then the force P in OA is zero and the conjugate is simply the force R itself.

(iii) *OA is a nul line.* I.e. OA lies in the nul plane of O. In fig. (i) of 14·7 BOA is then a straight line in the nul plane, and the components of R in directions OA, OB are both infinite, and the line OA is its own conjugate.

14·72. *To shew that the conjugates of all straight lines in the nul plane of a point O must pass through O.*

Let l be a line in the nul plane of O and m its conjugate. Any line OK in the nul plane of O is a nul line, but it meets l because l lies in the plane, and the conjugate forces in l and m have a zero moment about it, therefore it also meets m (rejecting the possibility of m being parallel to OK because of the arbitrariness of the direction of the latter). Therefore m meets every line through O in the nul plane; but m cannot lie in the nul plane (unless the system is coplanar, a case we need not consider), therefore m passes through O.

14·73. *To find the equations of the conjugate of a line whose equations are given.*

It is only necessary to choose any two points on the given line and write down the equations of their nul planes as in **14·63**. Then since these nul planes intersect in the conjugate of the given line their equations are the equations of the conjugate line.

14·74. Examples. (i) *Find the equations of the line conjugate to the line* $x = y = z$ *with regard to the force system* (X, Y, Z, L, M, N). *Find also the magnitudes of the forces in the line and its conjugate.*

The points $(0, 0, 0)$, $(1, 1, 1)$ both lie on the given line, and by **14·63** (1) the equations of their nul planes are

$$Lx + My + Nz = 0 \quad \dots\dots\dots\dots\dots\dots\dots(1),$$

and

$$(L - Z + Y)x + (M - X + Z)y + (N - Y + X)z = L + M + N.$$

Subtracting the first from the second, we have

$$(Y - Z)x + (Z - X)y + (X - Y)z = L + M + N \quad \dots\dots\dots(2),$$

and this with the first equation may be taken as the required equations of the conjugate line.

If P, P' denote the forces, the direction cosines of P are proportional to 1, 1, 1 and of P' are λ, μ, ν, where

$$\lambda : \mu : \nu = M(X - Y) - N(Z - X)$$
$$: N(Y - Z) - L(X - Y) : L(Z - X) - M(Y - Z).$$

Then

$$P/\sqrt{3} + \lambda P' = X, \quad P/\sqrt{3} + \mu P' = Y, \quad P/\sqrt{3} + \nu P' = Z,$$

so that

$$(\lambda - \mu) P' = X - Y.$$

This reduces to

$$(L + M + N) P' = \sqrt{\{\Sigma [M(X - Y) - N(Z - X)]^2\}},$$

and then we find that

$$(L + M + N) P = (LX + MY + NZ)\sqrt{3}.$$

(ii) *A system of forces is reduced to two forces one of which acts at a fixed point and lies in a fixed plane through the point. Prove that the other passes through another fixed point and lies in a fixed plane.*

Let P, P' be the two forces of which P acts at a fixed point O and lies in a fixed plane A. Since O is a point on P, the conjugate P' lies in the nul plane of O, i.e. a fixed plane. Again let O' be the nul point of the plane A; then by **14·72** since P lies in the nul plane of O' its conjugate P' passes through O', i.e. through a fixed point.

14·75. *The relation of conjugate lines to the central axis.* Let P, P' be a pair of conjugate forces and AA' the shortest distance between them. The central axis must intersect AA', for otherwise the system would have a moment about AA'; and since P, P' are both at right angles to AA', therefore R must also be at right angles to AA'. Let Oz be the central axis cutting AA' in O and making angles α, α' with P, P'. Let R, Γ be the force and couple of the wrench along Oz. Let $AO = a$, $OA' = a'$ and $AA' = h$.

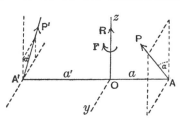

Then R may be replaced by forces Ra'/h at A and Ra/h at A', while Γ is equivalent to forces Γ/h at both A and A'. The forces at A must compound into P and those at A' into P', so that

$$R^2a'^2 + \Gamma^2 = P^2h^2 \quad \text{and} \quad R^2a^2 + \Gamma^2 = P'^2h^2.$$

Also $\tan\alpha = \Gamma/Ra'$ and $\tan\alpha' = \Gamma/Ra,$

or $\tan\alpha = p/a'$ and $\tan\alpha' = p/a,$

where p is the pitch of the resultant wrench.

14·8. Equilibrium of Four Forces.

If four non-intersecting forces are in equilibrium they act along generators of the same system of a hyperboloid.

Three non-intersecting straight lines determine a hyperboloid on which the three lines are generators of the same system. Consider the hyperboloid determined by the lines of action of three of the given forces. The algebraical sum of the moments about any generator of the opposite system is zero, but the three chosen forces intersect all generators of the opposite system, therefore the moment of the fourth force about every generator of the opposite system is zero; therefore it intersects all these generators and itself acts along a generator of the original system.

14·81. *Forces acting along generators of the same system of a hyperboloid are in equilibrium if they would be in equilibrium when acting at a point in the same directions.*

The equations of a generator of the hyperboloid

$$x^2/a^2 + y^2/b^2 - z^2/c^2 = 1$$

through the point $(a\cos\theta, b\sin\theta, 0)$ are

$$\frac{x - a\cos\theta}{a\sin\theta} = \frac{y - b\sin\theta}{-b\cos\theta} = \frac{z}{c}.$$

Let a force P act in this line—say at the point $(a\cos\theta,\ b\sin\theta,\ 0)$. Then if X, Y, Z, L, M, N are the corresponding components of force and couple at the origin

$$\frac{X}{a\sin\theta}=\frac{Y}{-b\cos\theta}=\frac{Z}{c};$$

and
$$L=b\sin\theta\,.\,Z=\frac{bc}{a}\,X,$$

$$M=-a\cos\theta\,.\,Z=\frac{ca}{b}\,Y,$$

$$N=a\cos\theta\,.\,Y-b\sin\theta\,.\,X=-\frac{ab}{c}\,Z.$$

For any number of such forces acting along generators of the same system the conditions of equilibrium are

$$\Sigma X=0,\ \ \Sigma Y=0,\ \ \Sigma Z=0\ \ \text{and}\ \ \Sigma L=0,\ \ \Sigma M=0,\ \ \Sigma N=0,$$

and it follows that if the first three are satisfied so also are the second three and hence the required result.

14·82. Examples. (i) *Shew that any system of forces acting along generators of the same system of a hyperboloid of one sheet can be reduced to two forces, one acting along a given generator of the same system and the other along some other generator.*

The system can be reduced to a force P along a given generator and a conjugate force P'. If we now reverse P and P' and combine them with the given system we get a system in equilibrium. Therefore the sum of the moments about any generator of the opposite system is zero. But this generator of the opposite system intersects all the given forces and the force $-P$, therefore it must also intersect $-P'$. Hence P' must intersect all generators of the opposite system and it therefore acts along a generator of the same system.

(ii) *Prove that any system of forces acting on a rigid body can be reduced to two, one of which acts in a given plane and the other is perpendicular to it. Also that (a) the shortest distance between the lines of action of the forces cuts the central axis; (b) if the point of intersection be fixed and the plane revolves about a line through the point perpendicular to the central axis, the lines of action of the forces generate a hyperbolic paraboloid; (c) the minimum value of the shortest distance is twice the pitch of the resultant wrench.* [T.]

Let the axis of the equivalent wrench be taken as axis of z and its intersection with the given plane as origin O. Choose axes of x and y so that the given plane is $z=x\tan\theta$. Then the resultant force R and couple Γ can be resolved into $R\sin\theta$, $\Gamma\sin\theta$, $R\cos\theta$, $\Gamma\cos\theta$, in directions OA, OB whereof the former is in the given plane and the latter normal to it.

The couple $\Gamma\sin\theta$ and the force $R\cos\theta$ compound into a single force $R\cos\theta$ parallel to OB and cutting Oy at E, where $OE=b$ is such that

$$\Gamma\sin\theta=bR\cos\theta.$$

Similarly the couple $\Gamma\cos\theta$ and the force $R\sin\theta$ compound into a single force $R\sin\theta$ cutting yO produced at F, where $OF = c$ is such that

$$\Gamma\cos\theta = cR\sin\theta.$$

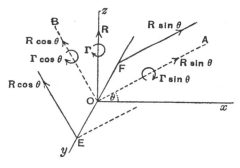

The system is thus reduced to a force $R\sin\theta$ at F in the given plane and a force $R\cos\theta$ at E perpendicular to it, and the line EF is perpendicular to both the forces and intersects the axis of the wrench.

Now suppose that the given plane revolves about Oy so that θ is variable.

The equations of the line of action of $R\cos\theta$ at E are

$$z = -x\cot\theta,$$

and

$$y = b = \frac{\Gamma}{R}\tan\theta;$$

so that the locus generated is $yz = -\dfrac{\Gamma}{R}x$, which is a hyperbolic paraboloid. Similarly the line of action of $R\sin\theta$ at F lies on the same surface.

Again, the shortest distance between the forces is

$$EF = b + c = \frac{\Gamma}{R}(\tan\theta + \cot\theta)$$

$$= 2\Gamma/R\sin 2\theta,$$

the least value of which is $2\Gamma/R$.

EXAMPLES

1. Two smooth planes intersect in a horizontal line and are inclined at the same angle α to the vertical. A uniform rod of weight W and length $2a$ is placed between them in a horizontal position making an angle θ with their line of intersection. Shew that the horizontal couple required to maintain equilibrium is $Wa\cos\theta\cot\alpha$.　　　　[I.]

2. Three smooth spheres of weight w and radius r rest symmetrically within a sphere of radius $3r$ and another smooth sphere of weight $3w$

is placed on the top of them. If the equilibrium is not disturbed, prove that the radius of the upper sphere cannot be less than $(\sqrt{2}-1)\,r$. [S.]

3. Forces act along the sides BA, BC, DA, DC of a tetrahedron, each force being inversely proportional to the side along which it acts; prove that if the four forces give a single resultant

$$AD.BC = AB.CD. \qquad \text{[S.]}$$

4. Three rods OA, OB, OC, each of length l and of equal weight, are smoothly jointed together at O and are placed symmetrically over a smooth sphere of radius a, the joint O being vertically above the centre of the sphere, and the rods resting against its surface. Prove that, if $\sqrt{2}l = 3a$, the rods, when in equilibrium, will be mutually at right angles to one another. [S.]

5. Four rods, each of length $2a$ and weight W, are smoothly jointed together and are placed over a smooth sphere of radius r ($r > a$) so as to be in equilibrium in the form of a horizontal square in contact with the sphere. Prove that the stress at each joint is equal to

$$\frac{aW}{\{2\,(r^2 - a^2)\}^{\frac{1}{2}}}. \qquad \text{[S.]}$$

6. $OABC$ is a tetrahedron formed by loosely jointing together six rods. A point X on OA is connected by a string, whose tension is T, to a point Y on BC. Shew that the action of the hinge at A on the rod AC lies along AC and that its magnitude is

$$\frac{OX}{OA}\frac{BY}{BC}\frac{AC}{XY}\,T. \qquad \text{[C.]}$$

7. A heavy sphere rests on three smooth pegs A, B, C in a horizontal plane. Prove that the pressures on the pegs are proportional to $\sin 2A$, $\sin 2B$, $\sin 2C$. [S.]

8. If A, B, C, D, A', B', C', D' be eight points in space so situated that $ABCD$, $A'B'C'D'$ are squares, then the condition of equilibrium of forces represented in all respects by AA', $B'B$, CC', $D'D$, when acting on a rigid body, is that the plane $ABCD$ should be parallel to the plane $A'B'C'D'$. [I.]

9. A weightless rod of length $2l$ rests in a given horizontal position with its ends on the curved surfaces of two horizontal smooth circular cylinders, each of radius a, which have their axes parallel and at a distance $2c$. The rod is acted on at its centre by a given force P and a couple. Find the couple when there is equilibrium and prove that the magnitude of the couple is least when P acts vertically, provided that

$$c < l \sin \phi + \frac{a}{\sqrt{2}} \sec \frac{\phi}{2},$$

where ϕ is the angle between the rod and the axes of the cylinders. [T.]

10. A heavy uniform ellipsoid is placed on three smooth pegs in the same horizontal plane, so that the pegs are at the ends of conjugate radii. Prove that the ellipsoid is in equilibrium, and that the pressures on the pegs are proportional to the areas of the conjugate central sections. [C.]

11. Wrenches of the same pitch p act along the edges of a regular tetrahedron $ABCD$ of side a. If the intensities of the wrenches along AB, DC are the same, and also those along BC, DA, and DB, CA, prove that the pitch of the equivalent wrench is $p + a/2\sqrt{2}$. [I.]

12. $ABCD$ is a regular tetrahedron. Wrenches of equal intensities act along BC, CA, AB, AD, BD, CD. Prove that the pitch of the resultant wrench is $\frac{1}{3}(p + q + r) + \frac{1}{4}a\sqrt{2}$, where p, q, r are the pitches of the wrenches which act along AD, BC; BD, AC; CD, AB and a is the length of an edge. [T.]

13. Prove that if equal forces act along the edges BC, CA, AB, DA, DB, DC of a regular tetrahedron, the central axis is the perpendicular from D to the plane ABC and the pitch of the equivalent wrench is $\dfrac{1}{2\sqrt{2}}\,a$, where a is an edge of the tetrahedron.

14. Shew that the following system of forces reduces to a single force:

Force $5\sqrt{2}$ acting through $(0, 0, \frac{1}{5})$ in direction $(3 : 4 : -5)$,

Force $3\sqrt{11}$ acting through $(\frac{1}{3}, 0, 0)$ in direction $(5 : 5 : -7)$,

Force $\sqrt{11}$ acting through $(0, \frac{1}{4}, 0)$ in direction $(-3 : 1 : 1)$,

the system being referred to rectangular axes. [I.]

15. A given system of forces is replaced by a force at a point P of a given plane and a couple. A line PP' is drawn parallel to the axis of the couple and proportional to its moment Shew that the locus of P' is a plane. [I.]

16. Forces X, Y, Z act along the three lines given by the equations

$$y = 0,\ z = c;\quad z = 0,\ x = a;\quad x = 0,\ y = b;$$

prove that pitch of the equivalent wrench is

$$(a\,YZ + b\,ZX + c\,XY)/(X^2 + Y^2 + Z^2).$$

If the wrench reduces to a single force, shew that the line of action of the force must lie on the hyperboloid

$$(x - a)(y - b)(z - c) - xyz = 0. \qquad [\text{T.}]$$

17. A given force acts along the axis of x and another given force along a generator of the cylinder $x^2 + y^2 = a^2$; prove that the locus of the central axis is an elliptic cylinder. [I.]

18. A given system of forces is equivalent to a wrench Γ of intensity R and pitch p: when an additional force R is added to the system the pitch of the resulting wrench Γ' is $p/2$. Prove that this additional force

intersects the axis of Γ and if the angle between them is 2ϕ, different from zero, then the axes of Γ and Γ'' are at a distance $\frac{1}{2}p\tan\phi$ and inclined at an angle ϕ. [I.]

19. $O\,A_1A_2A_3\ldots, O'A_1'A_2'A_3'\ldots$ are two straight lines in space and forces $\lambda_1.A_1A_1'$, $\lambda_2.A_2A_2'$, $\lambda_3.A_3A_3'\ldots$ act along A_1A_1', A_2A_2', $A_3A_3'\ldots$, respectively. Prove that they will be equivalent to a single force if $\Sigma\lambda_1.\lambda_2.A_1A_2.A_1'A_2'=0$; and to a single couple if $\Sigma\lambda_1=0$, $\Sigma\lambda_1.OA_1=0$, and $\Sigma\lambda_1.O'A_1'=0$. [I.]

20. A given system of forces is equivalent to a force and couple such that the angle between the axis of the couple and the line of action of the force is given. Find the locus of the line of action of the force. [I.]

21. Prove that a wrench is replaceable by forces on two lines of which one may be arbitrary and shew that if one line is at a given distance from the axis of the wrench the other is at a given inclination to the axis. [I.]

22. A system of forces can be reduced to a force R acting along a certain line l, and a couple G acting in a plane perpendicular to l. If the system is also reduced to two forces P and Q, such that the shortest distances between their lines of action and l are a, b, respectively, shew that
$$P^2(R^2a^2+G^2)=Q^2(R^2b^2+G^2).$$ [I.]

23. Prove that if three wrenches are in equilibrium their pitches, distances apart and inclinations are connected by the relations
$$p_1-z_{23}\cot\theta_{23}=p_2-z_{31}\cot\theta_{31}=p_3-z_{12}\cot\theta_{12}.$$ [I.]

24. Two wrenches of pitches p, q, whose axes are at a distance $2a$ from each other, have a resultant wrench of pitch ϖ, whose axis intersects the shortest distance between the axes of the given wrenches at a distance ξ from its middle point. Prove that the angle between the axes of the given wrenches is equal to
$$\tan^{-1}\frac{\xi(p-q)-a(2\varpi-p-q)}{\xi^2-a^2+(\varpi-p)(\varpi-q)}.$$ [I.]

25. Two wrenches have pitches p_1, p_2 and their axes intersect perpendicularly: the resultant wrench has pitch p and an axis which makes an angle θ with the axis of the first wrench and is distant h from that axis. Shew that
$$p=p_1\cos^2\theta+p_2\sin^2\theta,$$
$$h=(p_1\sim p_2)\sin\theta\cos\theta.$$ [T.]

26. The axes of two wrenches are along the axes of x and y, which are at right angles. The force and couple constituting the first are R_1 and G_1, and the second R_2 and G_2. Shew that the resultant wrench consists of a force $(R_1{}^2+R_2{}^2)^{\frac{1}{2}}$ and a couple $(R_1G_1+R_2G_2)/(R_1{}^2+R_2{}^2)^{\frac{1}{2}}$, and find its axis. [T.]

27. The axes of two wrenches are at right angles and the shortest distance between them is $2a$. Prove that the axis of the resultant wrench divides the shortest distance in the ratio

$$Q\{2aQ + (p-q)\,P\} : P\{2aP - (p-q)\,Q\},$$

where P and Q are the respective intensities of the wrenches, and p and q are the pitches. [C.]

28. Shew that two non-intersecting wrenches of pitches ρ, ρ' at right angles to one another can be replaced by two intersecting wrenches of pitches p, p' whose axes lie in a plane bisecting the shortest distance between the first two and also bisect the angles between them, and such that

$$2c^2 + 2c\,(p-p') = (p-\rho)\,(p'-\rho') + (p-\rho')\,(p'-\rho),$$

where $2c$ is the shortest distance between the first pair of wrenches. [I.]

29. Three forces act along given lines, no two of which meet or are parallel. Prove that if and only if the given lines are parallel to a plane, the magnitudes of the forces can be so chosen that the system is equivalent to a couple; and prove that the axis of the couple is in a fixed direction. [I.]

30. Prove that a wrench of intensity R and pitch ϖc may be replaced by two forces inclined at an angle 2θ to each other, the shortest distance between them being $2c$ and their magnitudes given by

$$\tfrac{1}{2}R\{\sqrt{(1 + \varpi\tan\theta)} \pm \sqrt{(1 - \varpi\cot\theta)}\}. \quad \text{[T.]}$$

31. Shew that if two conjugate lines meet any plane in A, A', then AA' passes through the nul point of the plane: also shew that if the lines are the common conjugate lines of two force systems their intersections with any plane and the nul points of the plane with regard to the systems form a range whose cross ratio is the same for all planes.
[T.]

32. Prove that the co-ordinates of the nul point of the plane $lx + my + nz = 1$, with respect to the system (X, Y, Z, L, M, N), are given by the equations

$$\frac{x}{X - nM + mN} = \frac{y}{Y - lN + nL} = \frac{z}{Z - mL + lM} = \frac{1}{lX + mY + nZ}. \quad \text{[I.]}$$

33. A system of forces is such that every tangent to the curve given by $x = a\cos\theta$, $y = a\sin\theta$, $z = b\theta$ is a nul line. Find the central axis and prove that the pitch of the equivalent wrench is a^2/b. [I.]

34. A sphere is described with its centre on the central axis of a system of forces. Shew that the nul points of tangent planes to the sphere lie on a hyperboloid of revolution. [I.]

35. Two equal forces act along generators of the same system of the hyperboloid $\dfrac{x^2+y^2}{a^2} - \dfrac{z^2}{b^2} = 1$, and cut the plane $z = 0$ at the extremities of perpendicular diameters of the circle $x^2 + y^2 = a^2$: shew that the pitch of the equivalent wrench is $\dfrac{a^2b}{a^2 + 2b^2}$. [I.]

36. Shew how to find the generators of the hyperboloids

$$x^2/a^2 + y^2/b^2 - z^2/c^2 = 1,$$

which are nul lines with respect to a given system of forces (X, Y, Z, L, M, N), and shew that these generators are parallel to one or other of the planes

$$Lx + My + Nz = \pm \left(\frac{bc}{a} Xx + \frac{ca}{b} Yy - \frac{ab}{c} Zz\right). \qquad \text{[T.]}$$

37. Forces P, Q, R act along any three mutually perpendicular generators of the same system of the surface $x^2 + y^2 = 2(z^2 + a^2)$, the positive direction of the forces being towards the same side of the plane xy. Prove that the pitch of the equivalent wrench is

$$2a\,(QR + RP + PQ)/(P^2 + Q^2 + R^2). \qquad \text{[T.]}$$

38. Any number of wrenches of the same pitch p act along generators of the same system of the hyperboloid $x^2/a^2 + y^2/b^2 - z^2/c^2 = 1$. Prove that they will be reducible to a single resultant provided their central axis is parallel to a generator of the cone

$$\left(p + \frac{bc}{a}\right) x^2 + \left(p + \frac{ca}{b}\right) y^2 + \left(p - \frac{ab}{c}\right) z^2 = 0. \qquad \text{[I.]}$$

ANSWERS

20. A circular cylinder.   33. Axis of z.

CAMBRIDGE: PRINTED BY W. LEWIS, M.A., AT THE UNIVERSITY PRESS

Printed in the United States
By Bookmasters